WIND SPEED

AN OVERVIEW

WIND SPEED

AN OVERVIEW

NICOLAS KOČÍ
EDITOR

nova
science publishers
New York

NOTICE TO THE READER

Library of Congress Cataloging-in-Publication Data

ISBN: 978-1-53618-412-9

Published by Nova Science Publishers, Inc. † New York

CONTENTS

PREFACE

In Wind Speed: An Overview, the history and development of wind energy is reviewed. Scientific trends in the academic field of wind energy are determined using a scientometric network analysis.

The relationship between wind speed forecasting and wind disasters is evaluated, particularly focusing on extra-tropical and tropical cyclones due to their dynamic origins.

Wind energy plays a significant role in clean energy sources, and the amount of energy that can be produced from a wind turbine is directly related to the value of the wind speed in that specific location.

The closing study focuses on wind as a source of energy in Kitka and Koznica, maintaining that in order to harness wind energy, it is necessary to carry out terrain condition analyses for the installation of wind turbines.

Chapter 1 - During recent decades the production and consumption of fossil fuels have caused a substantial increase in anthropogenic greenhouse gases. Since the mid-20th century, climate scientists have gathered extensive measurements of different weather patterns and associated environmental factors. These data indicate that Earth's climate has changed in almost every possible time scale since the beginning of the geological era and that the impact of human activities has been profoundly intertwined at least since the beginning of the Industrial Revolution.

Climate change is one of the most severe challenges faced in the 21st century affecting both the society and biodiversity. As these issues become more prominent each day, environmental awareness has increased consequently. Several international agreements imposing legally binding limits on greenhouse gas emissions to battle climate change and related environmental concerns worldwide have been negotiated. By the second half of this century, the energy transition is shaped towards transforming the global energy sector to a low carbon economy. Therefore, renewable energy sources such as wind, solar and biomass gained importance to mitigate carbon emissions to meet such criteria defined in these agreements for preventing further impacts.

Despite the fact all energy technologies have negative environmental impacts; renewable energy is relatively risk-free in comparison to those running on fossil fuels. Diversified sources of renewable energy were taken up as a commercial alternative to the production of fossil fuels in the last decade, rendering renewable energy projects a viable low-carbon alternative. In this manner, wind energy which had already been available non-commercially for several decades became even more preferable. The performance of wind energy systems has increased with the evolution of technology, thereby reducing costs and boosting the installation of wind turbines worldwide. Essentially in addition to the low levelized costs, wind-based capacities continue to grow rapidly, guided by the proven track record and financial stability provided by incentives.

In this chapter, the history and development of wind energy is reviewed. Main focus topics and scientific trends in the academic field for wind energy are determined using scientometric network analysis. The current global status is scrutinized in accordance with technical, social, political, environmental and economic aspects. For all above investigated aspects concerning wind energy, the five most important problems are defined and solutions to overcome these issues are proposed in line with literature findings. Finally, in light of these outcomes future projections are evaluated by logical framework for risks and opportunities.

Chapter 2 - Wind speed forecasting is one of the most relevant and challenging world research problems nowadays. It is related mostly to wind disaster forecasting, wind energy, coastal flood forecasting, other related issues. Several methods were developed for wind speed forecasting in different sectors and climates. The focus of this chapter will on the relationship between wind speed forecasting and wind disasters. There are a number of wind disasters based on the magnitude of wind speed and their origin. Wind speed and direction play an important role in regulating a wide variety of meteorological phenomena. A review of recent studies on accurate wind speed forecasting from different data sets, models, and algorithms will be the main aim of this chapter. This chapter also discussed the classification of extra-tropical and tropical cyclones because of their dynamic origins. Furthermore, the role of remote sensing in wind speed estimation was also reviewed.

Chapter 3 - The working principle of solar panels is the conversion of solar irradiation falling on the entire surface of the panel, to electricity. Conversion efficiency is acquired by capturing the electrical current generated when the sunlight interacts with silicon, multi-junction, GaAs, thin-film, perovskite or organic cells inside the solar panel. Although, this efficiency varies based on the specific features and size of the panels, the typical ratios are between 12 and 47.2% (NREL). All of these calculations are based on the estimation of the amount of solar energy from the sun under current seasonal conditions in order to achieve an accurate result and to obtain maximum efficiency by the solar panels.

Wind energy has taken an important place within the clean energy sources. The amount of energy that can be produced from a wind turbine is directly related to the value of the wind speed in that specific location. Therefore, it is very important to determine the wind speed accurately for each region. Besides that, with the transition to smart grids, the inclusion of different production resources into the grid has revealed the necessity of managing the grid efficiently.

In this study, data from the Modern-Era Retrospective analysis for Research and Applications, version-2 (MERRA-2) software [1] delivering time series of air temperature, local solar irradiation and wind speed were used for further calculations. For in situ measurement of the ground data, the measuring device was placed at an altitude of 50 m at the outdoor premises of the Institute of Solar Energy at Ege University, Izmir, Turkey. A statistical coherence among solar irradiation, air temperature and wind speed were observed and expressed as a third degree polynomial equation. With the equation developed, the differences between satellite data and ground data were corrected. Then, the differences between the corrected temperature and solar data were plotted. The value of the determination coefficient (R^2) in the resulting graphs has given the accuracy of the improved data. With this improvement, similar values have been achieved (approximately with 25% error for temperature and about 15% error for solar radiation and 0.0012% error for wind speed) using satellite data without ground measurement data. For this purpose, a different formulation was developed to obtain the results of a location-based measurement center from the satellite measurement results. This coherence obtained from solar irradation can be utilized to predict the changes over the years and perform calculations of e yield analysis.

Chapter 4 - Growing problems of global warming, environmental pollution and security of energy generation have increased interest in developing renewable and wind-friendly energy sources for different countries, and Kosovo has started it. This chapter studies wind as a source of energy in Kitka and Koznica. Initially, the study analyses the types of winds in the two potential sites in Kosovo, for which there are real measurements for the purpose of investing in this sector. Aerodynamics is also an important part of the study. In order to harness the wind energy, it is necessary to carry out terrain conditions analyses for the installation of wind turbines, which are supplemented by long-term correlation analyses. From the installed turbines, for their specific capacity and the number of potential places to put them, the annual electricity generation, generated by the wind park is calculated, taking into account the capacity factor of the wind turbines for this terrain. Further the economic analysis for wind park installation is given, where we can see what the lifecycle of two possible wind power generation projects can be. Economic analysis shows a life span of approximately 25 years for wind parks, and a return on investment of about 9.2 years, values that vary depending on the trend of selling a kWh, i.e., the price of electricity sold in a country.

In: Wind Speed: An Overview
Editor: Nicolas Kočí

ISBN: 978-1-53618-412-9
© 2020 Nova Science Publishers, Inc.

Chapter 1

A RETROSPECTIVE ANALYSIS OF WIND ENERGY

Dilvin Çebi[1], Fikret Müge Alptekin[1], Merve Uyan[1], Engin Deniz[2] and Melih Soner Çeliktaş[1,]*
[1]Solar Energy Institute, Ege University, Bornova, Izmir, Turkey
[2]The Beuth University of Applied Sciences Berlin, Berlin, Germany

ABSTRACT

During recent decades the production and consumption of fossil fuels have caused a substantial increase in anthropogenic greenhouse gases. Since the mid-20th century, climate scientists have gathered extensive measurements of different weather patterns and associated environmental factors. These data indicate that Earth's climate has changed in almost every possible time scale since the beginning of the geological era and that the impact of human activities has been profoundly intertwined at least since the beginning of the Industrial Revolution.

Climate change is one of the most severe challenges faced in the 21st century affecting both the society and biodiversity. As these issues become more prominent each day, environmental awareness has increased consequently. Several international agreements imposing legally binding limits on greenhouse gas emissions to battle climate

---

* Corresponding Author's Email: soner.celiktas@ege.edu.tr.

change and related environmental concerns worldwide have been negotiated. By the second half of this century, the energy transition is shaped towards transforming the global energy sector to a low carbon economy. Therefore, renewable energy sources such as wind, solar and biomass gained importance to mitigate carbon emissions to meet such criteria defined in these agreements for preventing further impacts.

Despite the fact all energy technologies have negative environmental impacts; renewable energy is relatively risk-free in comparison to those running on fossil fuels. Diversified sources of renewable energy were taken up as a commercial alternative to the production of fossil fuels in the last decade, rendering renewable energy projects a viable low-carbon alternative. In this manner, wind energy which had already been available non-commercially for several decades became even more preferable. The performance of wind energy systems has increased with the evolution of technology, thereby reducing costs and boosting the installation of wind turbines worldwide. Essentially in addition to the low levelized costs, wind-based capacities continue to grow rapidly, guided by the proven track record and financial stability provided by incentives.

In this chapter, the history and development of wind energy is reviewed. Main focus topics and scientific trends in the academic field for wind energy are determined using scientometric network analysis. The current global status is scrutinized in accordance with technical, social, political, environmental and economic aspects. For all above investigated aspects concerning wind energy, the five most important problems are defined and solutions to overcome these issues are proposed in line with literature findings. Finally, in light of these outcomes future projections are evaluated by logical framework for risks and opportunities.

Keywords: wind energy, scientometric analysis, logical framework matrix

INTRODUCTION

Wind power has been available at use for more than 3000 years and it has been used to generate electricity for more than a hundred years [1,2]. It had been used non-commercially for several decades until its technology started to develop. This development has always been associated with oil prices, as is the case in renewable energy technologies. During the 1970s oil crisis, technological advancements have first accelerated [3].

USA commenced with the research and development activities following the oil crisis, which lead to the evolution of wind energy from domestic use to wind farm installations. Due to the incentives issued by the US government, a rather rapid installation period was experienced in the 80s [2]. During this time the European wind market started to form. In the 90s, most of the market activity shifted to Europe [4]. In the last two decades, wind energy has become the far most prevalent technology worldwide in means of renewable energy, after hydropower [5].

According to Renewable Capacity Statistics published by IRENA (2019) [6] the pioneers in wind energy are currently China, USA and Germany based on installed capacities. Due to changes in energy policies towards a low carbon economy, an increased speed was observed in the last decade, enhancing the industry in other countries as well.

The urgency to enforce these policies became inevitable in a global sense, due to environmental, social and economic consequences of conventional energy systems, leading to the implementation of decarbonization scenarios [7]. The transition to a sustainable low carbon economy requires a global collective action in the attempt to prevent impacts of climate change and environmental pollution by promoting renewable energy technologies [8]. Emission reduction targets trigger the development and utilization of wind energy same as in other renewable energy technologies, increasing their share in the overall power sector [9]. Therefore, in line with the Kyoto Protocol and Paris Agreement, all types of renewable energy technologies are being developed in compliance with applied legislations and restrictions for energy [10].

Within this energy mix, wind energy plays an important role. The World Wind Energy Association announced the total capacity of wind turbines installed worldwide around 600 GW as of 2018. Due to stable installation of wind energy technologies, the competitive structure of the wind energy market and technological advances enabling offshore installations, wind energy is highly desired and widely used [5].

Transition to renewable energy systems not only means combating environmental consequences but also refers to the accessibility and security of energy, which should be scrutinized in all relevant aspects such as social, political, technical and economical. In this chapter, an analysis is executed for the academic studies performed on wind energy in order to identify scientific trend according to emerging topics. In line with the academic research, information on concerning patents is given and related aspects on wind energy such as technical, social and political, environmental and economic impacts are overviewed. For each related aspect the five most significant problems are identified and suggestions to overcome these impact inducing concepts are explained in accordance with literature.

WIND ENERGY IN ACADEMIC RESEARCH-SCIENTIFIC TRENDS

To maintain a current position in the energy field with a better understanding of its influencing factors, it is crucial to monitor and evaluate new information on developing technologies [11]. In order to gather recent information on the academic research done for wind energy, all publications within the last five years, from 2015 until 2019 are obtained using the Scopus search engine. The keywords used for the search are "wind energy" and "wind power". The total number of publications available in Scopus data base is 32,674. Relevant data suitable for further processing is limited in means of "citation information", "bibliographical information", "abstract and keyword" and "funding details" criteria available in Scopus engine.

In light of the web search, the big data of publications obtained from Scopus engine search are processed by the software program VOSviewer, developed for scientometric analysis by processing big data [12]. Filters on the software tool to further refine the data, making it specific to research on wind energy are utilized in order to see the trends and intensity in related research topics.

The VOSviewer mapping shown in Figure 1 focuses on related topics linked with wind turbines. It is seen that optimization and energy storage are the most weighted areas, representing a more concentrated research which should be anticipated due to the intermittent nature of wind energy which requires energy storage technologies to be developed and integrated to wind energy systems. Similarly, optimization is a current topic for operation and maintenance activities targeting high and stable power quality in relation to wind energy integration to power grids.

In regard to power quality of wind power plants, system components such as double fed induction generators (DFIG), wind energy conversion systems (WECS) and maximum power point tracking (MPPT) and relative technological advancements are of scientific interest as shown in Figure 2. Hybrid systems are also in focus which is consistent with the purpose of increasing system stability and power quality.

In Figure 3, it is possible to see the links between focus topics and other research areas which are interrelated. For instance, due to a technological development in a material or equipment optimization systems are also altered. The same applies for energy storage strategies and power quality issues that are shaped in parallel to technical advancements.

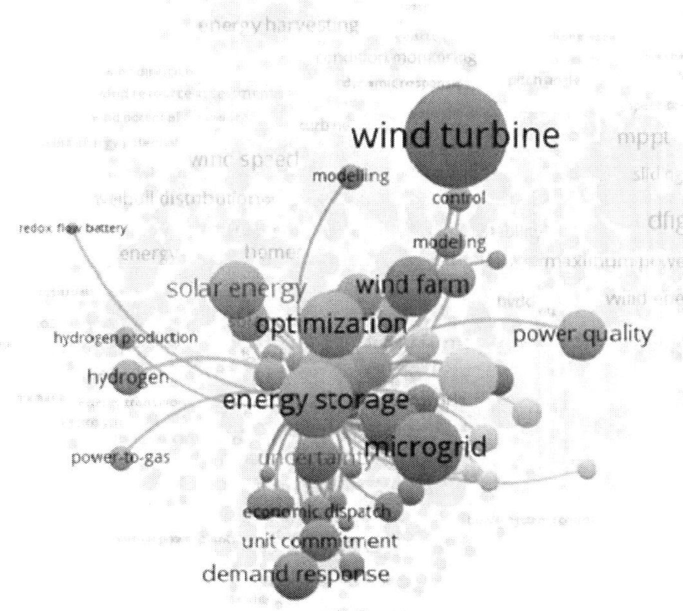

Figure 1. VOSviewer mapping for energy storage.

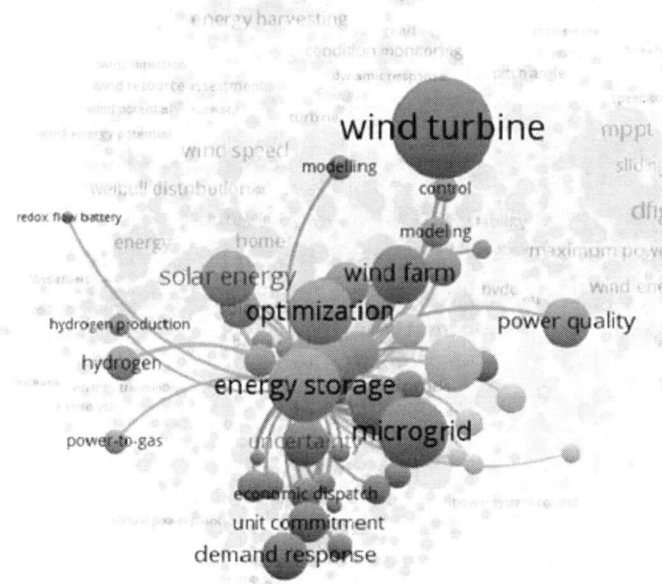

Figure 2. VOSviewer mapping for power quality.

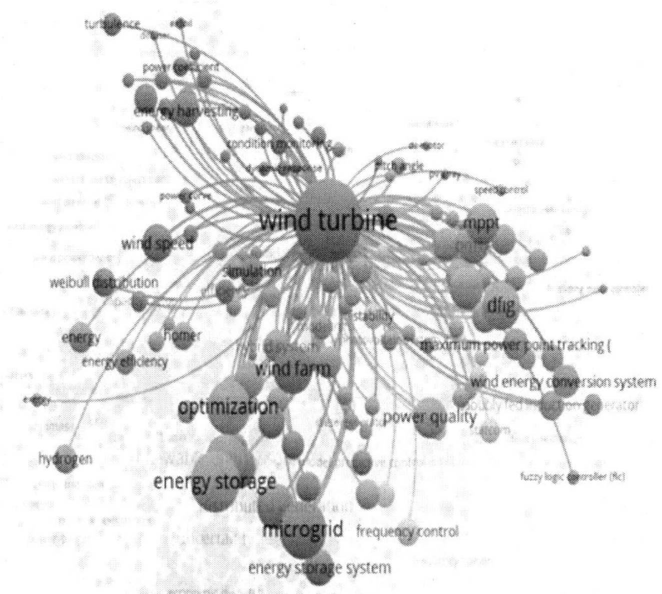

Figure 3. VOSviewer mapping for wind turbine.

In line with academic studies, patent applications should also be investigated to monitor new trends and developments. Patents are considered an important criterion for innovation, indicating the development process of technologies [13].

In the study conducted by Madvar et al., (2019) [14] patents published from 1973 to 2018 for wind energy are investigated. It is documented that Germany holds the most number of patents regarding wind technology. USA ranks the second, followed by Denmark and China.

The contribution of Asia to wind technologies is noteworthy, with China, Japan and South Korea having published the highest number of patents in the region. Japan is listed as the second best player after China. However, India is the second biggest electricity supplier using wind energy technologies. It is predicted that India will become the dominating force of the Asian market in the near future.

In the above mentioned study by Madvar et al. (2019)[14], sub-technologies of wind energy are evaluated. These are "generator or configuration", "blades and rotor", "control of turbines", "component and gearbox" and "nacelle". "Blades and rotor" and "control of turbines" are the two fields with highest potential, both referred to in more than 30% of published patents. It is reported that the "nacelle" domain had its growth stage in 2015, where "component and gearbox" is at the emerging stage. It is anticipated that "blades and rotor technology" will reach maturity in 2030.

According to the patent search conducted in this chapter, 684,563 results were found including the term "wind energy". Patent search was performed for the term "wind turbine" resulted in over 40,000 patents for the classifications F03D1/00/low, F03D7/00/low, F03D13/00/low, F03D9/00/low, F03D17/00/low. Identification of each class is represented in Table 1.

Based on the subject areas of focus regarding academic studies on wind energy, the categories can be classified as technical (including engineering, energy, computer science, material science, etc.), social and political (social sciences, arts and humanities, psychology), environmental (chemistry, earth and planetary sciences, agricultural and biological science) and economic (business management, economics, econometrics and finance). In the following sections of this chapter, the current global status of wind energy is examined according to technical, social, political, environmental and economic aspects.

Table 1. Patent classifications for wind turbin

Classification number	Identification
F03D1/00	Wind motors with rotation axis substantially parallel to the air flow entering the rotor
F03D7/00	Controlling wind motors
F03D13/00	Assembly, mounting or commissioning of wind motors; Arrangements specially adapted for transporting wind motor components
F03D9/00	Adaptations of wind motors for special use; Combinations of wind motors with apparatus driven thereby; Wind motors specially adapted for installation in particular locations (hybrid wind-photovoltaic energy systems for the generation of electric power)
F03D17/00	Monitoring or testing of wind motors, e.g., diagnostics (testing during commissioning of wind motors)

TECHNICAL ASPECTS

Wind power is one of the fastest growing sources and wind energy technologies are rapidly evolving. Wind power has the second largest share of renewable energies in the world after hydropower [15, 16].

The configuration of a wind turbine consists of a rotor with blades attached to a hub, a gearbox, shafts, support bearings, a tower generator and other electrical equipment. The materials used in wind turbine production and turbine sizes may vary [17]. The working principle is based on the conversion of kinetic energy of wind to kinetic energy by the rotor to rotational energy. Generated electricity by the generator can be directly used or fed to the grid. In order for wind energy to compete with other energy sources, turbine weight and cost are critical parameters. Rotor efficiency has been improved to a considerable extent by development and research activities [18, 19, 20].

Due to technological advances in wind power technologies in past decades, a significant increase in installed capacity of wind power plants is observed. While turbine technologies are developing, the need for cost effective measures in means of operation and maintenance is arising [21, 22, 23]. When social, economic and environmental aspects are considered for all energy sources, wind energy has relatively less negative impacts. By integrating wind power to energy systems at large scale, both emissions and operation costs will be reduced while sustaining a safer and cleaner production [24, 25, 26]. However, in order to secure system operation, problems causing negative impacts need to be assessed as well. These problems could vary from frequency and voltage fluctuations to balance problems for the grid caused by variable wind speed resulting in power losses[22, 24, 27, 28]. Therefore, it is urgent to resolve challenges of maintaining power quality at a desirable level in this integration.

Problems concerning the power quality of the systems occur as a result of the location and the discontinuous structure of wind power plants. The parameters for power quality are turbine type and design, characteristics and turbulence of the wind source, the location and positioning of the plant, grid and load properties, generator type and other electrical equipment [24, 29, 30].

Corrective measures are being applied for unforeseen downtime of wind energy plants. However, in order to establish a preventive strategy to minimize operation losses, prediction models focusing on reliability and failure are required [23, 31]. Building such models entail an extensive understanding of system operation and management. When wind power is integrated to the grid, the wind power output constitutes a problem for congestion management since the load cannot be constantly transmitted [32] and congestion in transmission networks causes several problems in system reliability [33]. In this regard, utilization of demand side management is an effective solution [34, 35].

Wind energy is considered as a clean energy source utilized to meet carbon emission reduction targets set due to increasing energy demand [36,

37]. However, the intermittent nature of wind power brings several challenges to the operation of wind energy plants [38, 39]. To overcome these challenges resulting from deviations caused by this intermittency, research activities for hybrid systems are being conducted [40, 41].

Xu et al. (2016) [42] conducted a technological paradigm research on the development of wind energy and suggested that the global annual installed capacity is anticipated to reach saturation by the year 2030. It is discussed in the same study that hybrid technologies (solar and wind) are more promising when compared to onshore and offshore wind energy technologies.

It is also noteworthy that in another attempt to reach higher nominal power in the wind turbines, the ongoing development shifted towards operating with better wind conditions. Following offshore implementations, efforts have been made to move to deep waters which led to the development of floating windmills [43].

TECHNICAL PROBLEMS AND PROPOSED SOLUTIONS

In order to integrate the wind turbine plants to electricity system, the plants must meet the electricity system requirements. The major challenges in integration are known as high cost and reliability problems of wind turbine plants when compared to existing commercial power plants. The technical parameters which affect system efficiency, operation and management are materials and structures; wind and turbulence[44], aerodynamics [45], control and system identification, electricity conversion, reliability and uncertainty modeling, design methods, hydrodynamics and soil characteristics [46].

The integration of several disciplines involved in the management of wind energy systems to design methods is an ongoing issue in design optimization. Optimization aims to achieve maximum efficiency with minimum hindrance of operation requirements in accordance with power quality requirements at maximum reliability. Power curve models used to achieve maximum efficiency does not include the dynamic behavior of conversion processes [47]. For an accurate optimization information on both complex processes should be analyzed.

Available prediction methods today do not involve the intrinsic failure mechanisms of materials and components [48]. Existing technologies and software tools are used for structural health and condition monitoring [46, 49, 50], but the correlation between damage state and service life cannot be fully explained. The behavior of several materials used for wind turbines, especially that of composite materials is not known to a broader extent [46]. With further research on behavior characteristics of materials and their interactions with each other, these findings could be integrated into prediction models leading to more accurate results and consequently higher performance [51].

Wind turbine controllers are desired to maximize the energy capture within the turbine's operational limits. Integration of weather prediction data, distributed sensing, fatigue load predictions, related information on operation and management, grid requirements and other control aspects into models designed for minimizing fatigue loads while maximizing production is considered to increase plant production while preventing failures occurring due to unforeseen conditions [52].

For the wind turbines to be installed at optimal locations, wind conditions should be thoroughly known, in relation with precise information on aerodynamic and atmospheric aspects[53, 54]. Although generic equations are known for these aspects, exact modeling for airflow is not possible since these equations are nonlinear and not location specific [46]. Combining different models in a way to achieve best possible data is one solution to overcome this problem.

Small scale wind turbine systems also have some deficiencies as large scale systems. The main problems of small scale turbines are high cost and low reliability [55, 56, 57].

Concentration on new components and turbine development for sites with low wind speed is inevitable, as well as research on parameters concerning reliability issues such as lightning, corrosion, bearing lubrication and insulation[58]. Further research on durability, blade aerodynamics, power electronics and noise elimination which reduces productivity of the system may help overcome technical hindrance of small scale wind turbines. Integration with other distributed generation and energy storage technologies is required [59, 60, 61, 62, 63, 64].

Among all above mentioned technical issues, noise elimination presents itself as an urgent matter since noise is one of the most widely vocalized concerns of wind power plants, triggering all related aspects investigated in this chapter. Therefore, it is critical for research to accelerate towards reducing and/or eliminating noise at the wind turbine design stage.

SOCIAL AND POLITICAL ASPECTS

All technological shifts in history are shaped by social, political and economic parameters affected by society and energy projects are no exception. Wind energy as a renewable energy technology should not be evaluated solely as a technical endeavor but as a socio-technical structure. It is important to understand that whether a technology fails or thrives is dependent on a complex nature of several parameters [65, 66].

When the policy programs regarding renewable energies first commenced in the eighties, social acceptance was not regarded among primary concerns[67], since primary surveys showed a high public acceptance rate especially for wind power. The same attitude continued to present itself in the 90s due to high level of support for renewable energies.

When it comes to renewable energies the issue of social acceptance should be examined through a different view. Renewable energy plants are usually at a smaller scale than conventional plants, thus requiring a higher number of siting decisions. Renewable energies are also more visible in means of energy harvesting which makes them more of an obvious concern for those who interact closely with plant implementations [66]. Additionally, since renewable energy technologies cannot currently compete with conventional energy technologies, their acceptance becomes a choice between short-term costs and long-term benefits.

The decrease in social acceptance of wind energy plants can be attributed to the fact that social implications had been neglected for several decades. Consequently, social acceptance status has a limiting effect on the wind energy industry [68]. Relevant parties in wind energy industry make efforts to improve social acceptance by enhanced public participation programs, new policies and ownership regulations and community benefit schemes [66]. Public opinion polls and other indicators show that the public support for renewable energy technologies is high in many countries, even when government support is insufficient.

Factors obstructing successful project implementations for renewable energies occur due to lack of social acceptance [69] that is identified with three dimensions as socio-political, community and market acceptance [70].

Parameters effecting socio-political acceptance can be listed as ownership issues, economic situation, high electricity prices, technologies requiring high investment costs, difficulties in credit and financing and insufficient and/or non-existing regulatory standards [71].

Public opinions are influenced by factors such as mistrust in investors [72], lack of information on proposed technologies and lack of impartial perception [73]. It is experienced that a comprehensive system including stakeholders and local politicians have a positive impact on social acceptance. Policies which mandate efficient processes, emission reduction targets, referring to both local and national energy policies, support through incentives and license procedure simplifications are essential.

Wolsink (2007) [70], Dimitropoulos and Kontoleon (2009)[74] specified two main institutional factors as procedural justice and distributional justice. In order to assess distributional justice for wind energy projects, an understanding of financial participation on social acceptance is essential [75].

Lack of social acceptance emerges as a result of "Not In My Backyard" phenomenon [76]. Communities tend to agree with the projects only if they are not implemented in their "backyard". For wind energy projects however, this view does not fully explain communities' opposition. It is shown that acceptance levels can be increased with proper involvement of the community [52].

IEA principles state that acceptance levels of communities increase with their perception of an inclusive, respectful, transparent and consistent process which is sensitive to local values. A satisfactory discussion procedure should also be included in the process [77]. Towards governments who establish the required conditions to maintain procedural and distributional justice are proven to be more trustworthy and receive higher levels of social acceptance in proposed projects [77]. Walker et al. (2014) [78] pointed out that trust should be considered as a complex issue in renewable energy projects especially when "community" should not be accepted as a singular body.

Sposato and Hampl (2018) [79] reported that the acceptance trend of renewable energy projects can be demonstrated in accordance with a u-shaped curve, which indicates high acceptance prior project development to lower rates during planning and siting, increasing after completion.

Wilson and Dyke (2016) [80] studied the acceptance status of a wind energy project for 5 years including periods both before and after project implementation, which revealed the u-shape curve as well. It is suggested that high acceptance levels towards wind energy projects due to support for renewable energy does not necessarily mean stable positive opinion of residents as they engage and live with the projects.

Management procedure should be sustained before and after construction of the wind power plant. The management carried out after the construction is defined as "adaptable management"[81]. While it constitutes a risk factor for the investor due to financial consequences, it is beneficial for preventing environmental impacts.

ENERGY TRANSITION TO LOW-CARBON ECONOMY

Renewable energy plays a critical role in the transition to low-carbon economies. Therefore, several regulations are enforced in many countries. Environmental regulations are being applied especially in industrialized countries for electricity generation towards energy transition, which resulted in policies incentivizing renewable energy sources. Most common models are feed-in-tariff (FIT) and tradable certificate systems. Feed-in-tariff is characterized by the obligatory connection to electricity from renewable energy sources of network operators and subsidization of producers for higher costs of electricity from these sources [82]. Tradable Green Certificates are granted to electricity producers form renewable energies and are processed according to the market conditions [83].

FIT is a support mechanism used to encourage renewable energy technology investments [84]. According to the investigation conducted by Ringel (2006) [85] on these support mechanisms in the EU, it is shown that FITs are an effective tool to promote and increase renewable energy use by individuals. Additionally, carbon taxes practiced by governments impose higher costs for conventional electricity generation while incentivizing renewable energy investments.

In Germany, although the public support was quite high following the introduction of a FIT for renewable energy, objections are being vocalized as a result of increasing costs both for households and industry. However, when assessed globally, Khorsand et al. (2015) [77] stated that social acceptance is favorable for wind energy since it is perceived as an ethical and relatively safe renewable technology.

SOCIAL AND POLITICAL PROBLEMS AND PROPOSED SOLUTIONS

Based on above review of social and political aspects concerning wind energy, main problems are given below along with proposed solutions.

One of the challenges to be dealt with under social acceptance is distributional acceptance which is related to the financial participation of the community. This challenge can be solved by offering financial contribution to communities within the context of wind energy plan [86]. This procedure has already been applied by such countries like Germany and Denmark. Germany passed the legislation for the financial participation of communities in renewable energy projects. In Mecklenburg-West Pomerania in newly planned wind energy projects, the developers are obliged to offer financial participation to all communities residing within a 5 km radius of the project site [75]. It is suggested that such participation rights enables communities to be involved in wind energy projects and profit financially without taking an active role in project implementation [75].

The other problem in means of social acceptance is procedural acceptance. To overcome this problem, the community should be informed about consultation, planning and involvement. IEA proposes a categorization for the parameters influencing social acceptance as: policy and strategy, well-being and quality of life, costs and benefits, consultation and involvement and implementation strategy [87]. While maintaining local involvement in new projects, it is also important to find out the views of those who are impartial [81]. To increase attendance to informatory meetings and improve public consultation, the "consensus-building process" needs to be enhanced.

Another important issue related with social well-fare is the negative impacts of wind energy on human life. The most notable impacts which cause discomfort to communities in this sense are visual impacts and noise. It is also experienced that the turbine equipment interferes with electronic equipment used by the community. Considering that the visual impact is influenced by shape, color and positioning of the turbines, the distance from the residential areas, flickering and location of the plant, it is suggested that uniform row of wind turbines at same size with light colors may help ease the disturbance of the nearby communities.

Windiness of plant locations is the most important factor for investors in the construction of wind farms. Different FIT mechanisms can be applied depending on the wind plant location for increasing investments. A good practice is the Renewable Energy Law of Germany put into force in 2000, where the FIT levels are arranged as to encourage implementations in less windy locations, while maintaining the renewable energy support with a 20 year FIT [88, 89].

One of the most significant issues regarding the political aspects of wind energy is connected with the financial pressure imposed by subsidies for renewable energies. Renewable energies are being developed and promoted due to their potential in meeting the energy demand while reducing carbon emissions when compared to conventional energy generation technologies. However, the intermittent nature of renewable energy sources, lack of technological development and high investment costs make it difficult for these technologies to compete within the electricity market. Renewable energy incentives and subsidies are being applied to increase the share of renewables in the energy mix. There has been intensive research to encourage renewable energy development accompanied by carbon emissions reduction, while reducing the financial pressure on governments[90, 91, 92, 93]. A tradable green certificate mechanism is proposed and adopted by many countries as a way to release financial pressure [94, 95].

ENVIRONMENTAL ASPECTS

In order to reduce the environmental footprint of energy generation systems based on fossil fuels, renewable energy systems are becoming more widely installed globally. As of 2017, 70% of net additional power generating capacity was from renewable energy sources, where a quarter of that energy was from wind power [96].

However, systems running on renewable energy sources are not totally neutral in means of environmental impacts due to energy and materials required during their production, installation and even dismantling stages, although they are considered to be a cleaner source of energy. In this manner Life Cycle Assessment (LCA) used for the impact assessment of a process or product for its lifetime [97] is considered to be a highly useful tool to evaluate the environmental footprint of renewable energy systems to conventional technologies [96] calculated as direct or indirect carbon dioxide emissions throughout the lifetime of energy systems. LCA evaluates the technologies for energy production in association with primary energy consumption and greenhouse gas (GHG) emissions. To accurately determine the carbon footprint of a system upstream and downstream processes are examined [98].

Based on LCA studies [99, 100, 101], wind turbines with higher hubs are proven to have a lesser environmental impact and horizontal axis wind turbines are found to be more preferable in means of energy intensity than vertical axis wind turbines than [97, 101]. Life cycle assessments conducted by Weinzettel et al. (2009) [43] and Wang et al. (2019) [102] suggested that alterations in transport operations of both onshore and offshore wind turbines could result in reduced emissions considering that transport amounts to a large quantity of overall GHG emissions of wind power plants.

Although CO_2 emissions during wind power plant operations are negligible with respect to energy generation technologies using non-renewable energy sources[103], fossil fuels involved in its application contribute to the carbon footprint of wind power plants. Both onshore and offshore wind power technologies operate with low emissions however their manufacturing, raw material extraction, operation and final disposal require energy consuming activities that all result in GHG emissions [98, 104].

Due to environmental cost of conventional fossil fuel based energy production, environmental policies proposed to emphasize on renewable energy power in order to prevent environmental problems such as global warming and air pollution[105]. Basically there are two types of environmental policies which are carbon taxes and tradable permits. The European Union (EU) member countries, the United States, Italy and New Zealand prefer tradable permit policy while carbon tax has limited utilization in countries such as Sweden, Finland, Ireland, Norway and Denmark, including some provinces in Canada and the United States [106]. The EU carbon market, the Chicago Climate Exchange, and the Australian Stock Exchange in New South Wales are three most important carbon markets.

The largest carbon trading market is The European Union Emissions Trading System (EU ETS) which was founded in 2005 by the European

Union. EU ETS has led to the launch of carbon markets in other countries while creating new opportunities for investors [105, 107]. As a result of these actions, many countries have been involved in carbon exchange. The carbon trading scheme of the largest carbon emitting country China was initiated in 2017 [108] and eight carbon emission trading markets have been launched since 2013 [109].

ENVIRONMENTAL IMPACTS OF OFFSHORE WIND POWER PLANTS

Offshore wind energy has superior qualities in means of wind capture than onshore wind energy. Offshore wind plants require higher investment and operation costs compared to onshore plants[110, 111]. However, considering continuous wind and higher speed enabling full-load operation as well as less turbulence [112] increasing turbine lifetime compensate for these costs, making offshore plants an economically attractive option [113]. However, research has revealed that properties of offshore wind plants such as distance from the land, the size of the plant and turbine height have significant environmental impacts [111, 114, 115].

The marine environment is subject to severe impacts at all stages during the lifetime of offshore wind plants. For instance, even the decommissioning of the offshore wind plants is of consideration due to the foundation of the plant installed as an "artificial coral reef" allowing new habitat to grow [116]. Alteration or destruction of the surrounding marine habitat may constitute a risk if the foundation is totally removed at the end of the lifetime of the plant [116].

During construction and operation of offshore wind plants noise is an important issue on marine animals which are sensitive to noise, causing changes in their behavior [117]. Moreover, magneto-sensitive species can be affected by electromagnetic fields generated by the underwater cables [116, 117]. Marine animals communicating through acoustic signals may be affected by the noise generated in offshore wind power plants, as well as other fish sensitive to loud sounds [118]

Offshore wind power plants have higher negative impacts on local bird populations [119] compared to onshore plants due to their height and size [120]. Offshore wind plants may cause an obstruction on migration paths and the noise may inhibit the acoustic perception of bird populations. On a different note, Gasparatos et al. (2017) [121] points out an indirect positive effect for animals on land of which the survival rates may be affected by the decline in bird populations.

Apart from ecological aspects and stress on the marine environment, sea and sea food quality might be severely affected from chemical emission during construction and operation of offshore wind power plants [113]. Corrosion protection systems are one of the sources of chemical emission in offshore wind plants, protecting the steel construction from corrosion and supporting the stability of the plant [113, 122]. Epoxied resins and polyurethane based coating widely used in marine applications to protect the steel structure from corrosion may release chemical emissions to seawater especially from areas of the power plant in direct contact with water[113, 123].

Another protection method antifouling paint [124] is applied on the surface of offshore wind turbine plants to prevent corrosion or accumulation of life forms on the applied area. Diffusion of antifouling paint may be harmful to marine life due to the chemicals in the paint [111].

When considering environmental and ecological effects, both wildlife and human life are affected by wind energy. One of the factors affecting the opinion of societies is the visual impact of wind power plants, which is considered among the environmental impacts due to its influence on human life. According to Sklenicka and Zouhar (2018) [125], when the aesthetic aspect of a landscape is viewed as a combination of subjective component such as the psychological perception of people and subjective components referring to the visual properties of the land, both aspects influence the acceptance of wind turbines consequently leading to rejection or approval of wind projects. Therefore, the objective component and the subjective element of the assessment need to be evaluated prior implementation.

ENVIRONMENTAL PROBLEMS AND PROPOSED SOLUTIONS

Due to implementation of renewable energy systems, clean and low emission technologies are being developed to mitigate climate change and reduce environmental impacts. However, these technologies are not totally carbon-neutral. In order to eliminate negative environmental impacts of these technologies, eco-friendly products should be developed at the design stage [97] where 80% of the environmental impact of a product or service is determined [126]. Integrating LCA approach to the design and planning of products can eliminate emissions and waste generated while reducing manufacturing costs.

Both onshore and offshore wind power plants have a serious noise issue both during construction and operation stages. Shafiullah et al. (2013) [24] stated that the mechanical noise could be prevented at the designing phase by the application of insulation inside of the turbine housing and aerodynamic noise could be reduced by blade design. The impact of noise generated by offshore wind power plants on marine ecosystems due to impact pile driving [127] can be avoided at a certain extent [128]. Development of alternative systems for noise mitigation such as isolation casing, vibratory piling and gravity base foundation techniques are required [129].

Chemicals used for corrosion protection in offshore wind power plants can damage marine life by the diffusion of chemicals to seawater [113]. The offshore wind technology is relatively new so the effects on marine life are not completely known. Emissions resulting from wind power plants are of relatively low-risk when compared to other chemical pollutants posing a threat to marine environment. Currently, the actual data gathered from offshore plants is not comprehensive enough to assess the actual impact. Relative analyses should be done in the surrounding environment of offshore wind power plants. Kirchgeorg et al. (2018) [113] also proposed that biomarkers for gas platforms could be used for monitoring purposes.

Wind power plants installed on migration routes of bird populations is another issue, although it is relatively negligible since other urbanization activities cause even more damage to bird species [24]. This problem is also avoidable with necessary precautions taken at the design stage of wind power plants. It is known that turbines with low hub heights have a higher potential of harming birds due to the high rotation speed of blades [100, 115. The nesting of birds can also be prevented by using smooth tubular steel towers[110, 130]. Demir and Taskin (2013) [100] also state that turbine location is of significance in means of environmental impact due to installations in locations with optimum wind speed.

In line with low emission targets according to decarbonization scenarios, uncertainty of carbon price is one of the environmental problems in this context as in all renewable energy systems. Regulations towards an effective price forecast are necessary to maintain the stability and sustainability of the carbon market [105, 172].

ECONOMIC ASPECTS

The success or failure of any investment into renewable energy sources depends quite substantially on the improvements in technology and market regulations. Hardly any other investment relies as heavily on the availability of renewable resources such as wind energy as this type of investment does, with the investor having close to no power over manipulating it in his favor, if necessary. For any investor, a favorable market for an investment is thus a rare and highly coveted asset which is to be protected against rival intervention for the entire duration of a project.

In order to meet increasing energy demand, renewable energy resources and the technologies based on conversion of those resources into energy are drawing attention in the world through industrialization as part of measures against global warming and climate change and due to the fact that fossil resources are exhaustible, nuclear energy generation technology is held by developed countries. Cost increases in fossil fuels and their environmental impacts have made the renewable energy a strategic business field, and the use of biomass (biofuel), hydraulic energy, wind energy, solar, geothermal energy and other renewable energy technologies became widespread.

Today, all renewable energy resources are meeting only almost 4% of the energy demand whereas the projections prepared by International Energy Association (IEA) anticipate that renewable energy sources (RES) should meet 12.4% of the total demand by 2023, and an investment of 10.5 trillion dollars is expected to be made on renewable energy resources between 2001-2030. In the same period, RES share in energy generation among OECD countries is expected to reach 25%. Requirement for reduction of carbon dioxide ratios, ensuring energy supply reliability in countries dependent on fossil fuels and expectations regarding RES's cost advantages in medium and long term compared to fossil fuels helped getting investments and supports on RES. EU council is also focusing their energy policies on improvement of RES such as wind, solar, biomass and hydraulic energy in particular. EU aims to increase energy consumption from renewable energy resources which is currently at 6% level to two fold [132]. EU is placing utmost attention on renewable energy in order to reduce both their dependency on fossil fuels and greenhouse gas emissions. European Union focuses on electric, biofuels, heating and cooling fields in order to make 32% of its general energy portfolio from renewable energy by 2030 [133].

With growing populations and quickly increasing economies, especially developing countries are beneath immense pressure to fulfill the growing demand for electricity and to produce electricity to all. One of the problem is move to a new energy market focused on renewable manufacturing technologies while governments planify the develop plan in order to regulate energy market towards renewable energy.

The wind energy allows for a significant realization of energy supply faster than any other technology which has been demonstrated worldwide by developments in the industry during the past decade. Additionally wind energy has undergone immense innovation and growth over the past decade, driven mainly by policy and technological improvements [134]. As seen in Figure 4 derived from the data published by International Renewable Energy Agency (IRENA) [135], the levelized cost of energy

(LCOE) of onshore wind fell 34% between 2010-2018 from 0.084 USD/kWh to 0.055 USD/kWh, making it increasingly competitive at utility scale while the LCOE of offshore wind has a significant cost reduction of 20% between 2010-2017 [136]. The LCOE of photovoltaic (PV) plant has changed tremendously over the past decade from 0.370 USD/kWh to 0.085 USD/kWh which brings competitiveness with wind energy in the market. The LCOE for a wind power plant is the ratio among the total fees of the plant and its total energy production over its monetary lifetime. The LCOE for wind and solar energy plants are based often on purchase investments to assembly and installation the plant; financing conditions (go back on investment, interest, plant lifetime); operating prices over the lifetime of the plant (insurance, maintenance, repairs) and lifetime and the annual degradation of the electricity plant [137, 138].

Renewable energy technology costs reducing dramatically shows a falling tendency since decade. The most cost-effective source of power for many energy markets and economies has been wind and solar power, and rates are set to continue reducing in next years.

Advanced technologies for renewable energy generation, network deployment, advanced methods of anticipating and improvements to wholesale market legislation, and other policies encouraging the creation of renewable energies have promoted tremendous growth in world wind power [139]. Number of growing wind power plants not only provide reliable energy source and comprehensive approach of climate-friendly, yet due to the reduction at the cost of production on the energy market is also becoming more competitive.

Wind energy is the most rapidly expanding source of energy in the world and wind power is one of today's most commonly used alternative energy sources. As shown in Figure 5 derived from The Global Wind Energy Council statistics [140], the global offshore wind industry made a major step forward in 2015, with over 3.4 GW installed in five markets worldwide and total offshore wind capacity reaching more than 18.8 GW in 2019 [141, 142].

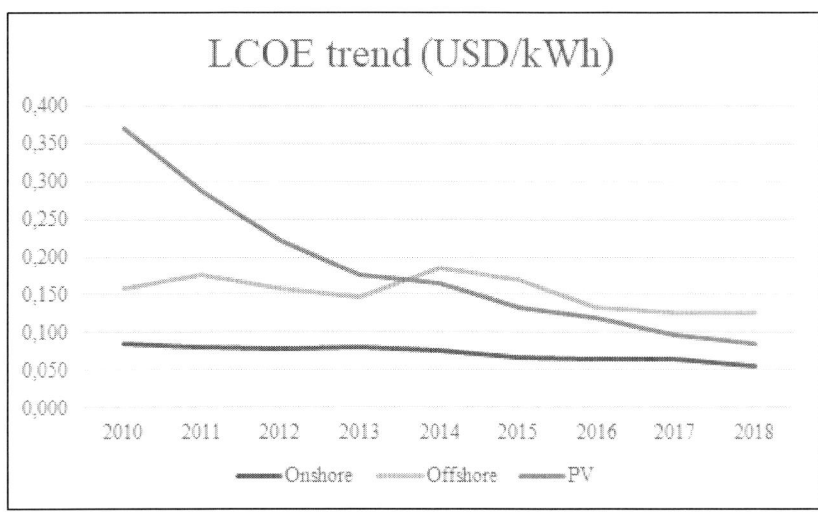

Figure 4. Global weighted average LCOE for onshore wind, 2010–2018.

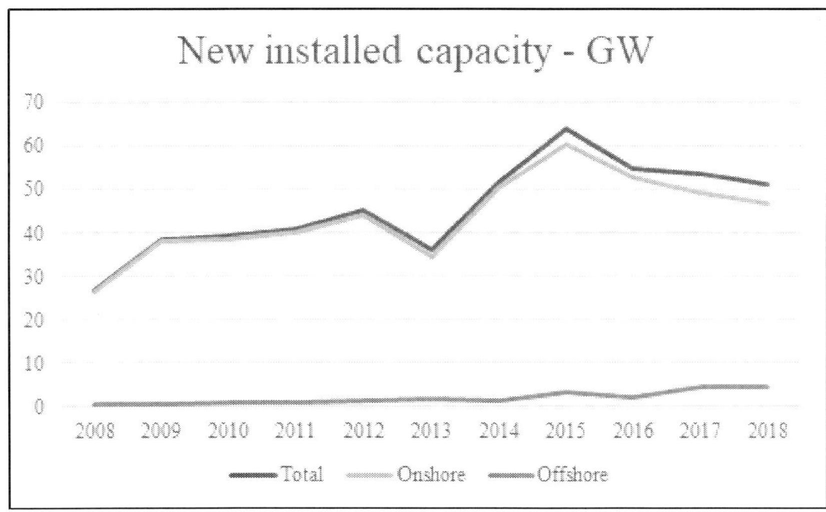

Figure 5. New installed capacity of onshore and offshore wind energy power plants by year.

China, which accounts for around 6.4% of the global market, is the largest market outside European waters. But other countries set aggressive wind market development goals, and some of these markets are beginning to take off [142, 143]. Figure 6 produced based on statistical data of IRENA [144] shows the cumulative capacity of installed wind power worldwide. World's cumulative installed onshore wind capacity at the end of 2018 reached 597 GW, across a total of 3,589 wind turbines added in 2018 [141].

The USA's target is to use wind energy increasingly to make it competitive in the interim energy market by 2035 and reduce system and production costs through use of advanced technologies for competition in the main energy market by 2050. The cumulative capacity of installed wind power increased by 8% between 2017-2018, half of which was installed in Asia-Pacific.

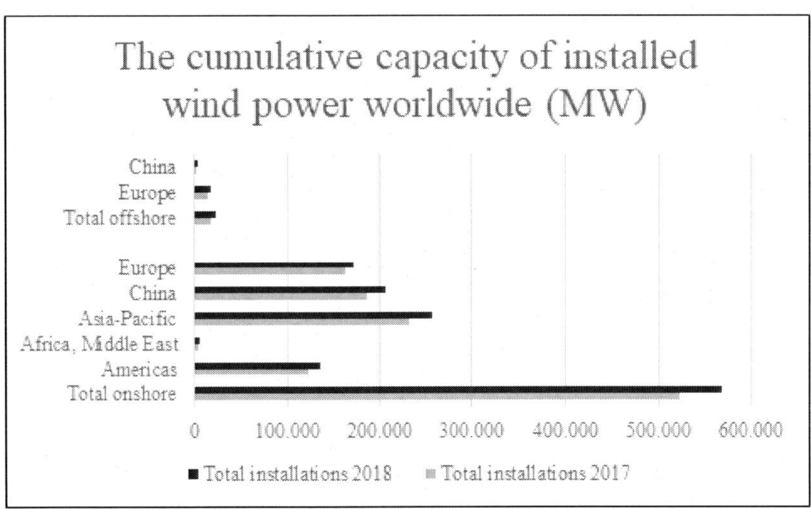

Figure 6. The cumulative capacity of installed wind power worldwide.

Energy business is a strategic field qualified as a vital importance for countries' development policies. Population increase, industrialization and urbanization concepts around the world and the commercial opportunities increasing due to globalization increase the demand for natural resources and energy every day. Projections of the IEA depict that in case existing energy policies and energy demand preferences are continued, the world's primary energy demand will increase in 40% until 2035. World's primary energy demand in this circumstance which is referred as reference scenario shall increase from 16.07 billion TEP (ton equivalent petroleum) level in 2018 to 18.05 billion TEP in year 2035 [145].

It is in Asian countries that the renewables industry branches out rapidly and generates thousands or even millions of workers. Ultimately, wind energy development is increasing dramatically at a low cost, with China and Japan based. In the fields of renewable energy use and employment growth, China, Pakistan, India, Indonesia, Japan, France, Bangladesh and Colombia are global leaders in the field as well. Germany still has the highest number of jobs in renewable energy, by far, in Europe. That's double that of France and ahead of other Member States, including Italy or Spain. Recently Germany set its target of 65% renewable energy by 2030 [146].

Support policies on renewable energy, innovation, growth and grants for consumers, tax and technical rewards for buyers, quality and quantity support programs, such as tariff benefit assurances, net assessment and renewable energy, generic procurement list, licenses and requirements for renewable energy. The FIT schemes for renewable power services for electrical systems are valid 20 years in advance. Feeding premiums give an excess of the wholesale energy market price a fixed premium guarantee [147].

Results of both investment profitability and capital profitability in energy investments are agreed to provide positive outputs in terms exceeding 10 years around the world. Common characteristics of investments in energy business is that although investment return is long term, they create a sustainable market environment and profitability rate per unit product is very high.

Most of the operation of the wind industry also takes place on property. China has led the way for the first time with USD 11.4 billion in offshore ventures. The budget of USD 3.3 billion was earmarked for European programs [148].

Ultimately, over one million people work together onshore and offshore wind power plants internationally. In a limited number of countries, the bulk of wind employment are located, although the concentration is smaller than that in the PV market. China represents 44% of the total worldwide; 75% of the top five countries. Also in the regional environment the solar PV industry is more balanced.

Figure 7 produced based on the data obtained from IRENA [149] and Frankfurt School-UNEP Centre [150] reports representing the industry trends and prospects since 2004, reveals that global renewables expenditure in 2018 exceeded US$ 272.9 billion, well outstripping new fossil fuel generation investment. In 2018, renewables investments exceeding 250 billion dollars were the fifth year in a row. Renewable energy is becoming cheaper, particularly wind energy.

The lower on-shore wind electricity costs of 2018 were powered by ongoing decline and increases in the average power factor in total cost deployed. Continuing advances in the production of turbines, increasingly flexible global supply chains and a growing number of LCOE turbines in a variety of operational conditions all fuel this development.

Figure 8 shows the price indices and trends of wind turbines between 1997 and 2018 [151]. Since its peak in 2007–2010, wind turbine prices have decreased from 44 to 64 percent, depending on the market. Since 1998, the price of the Chinese wind turbine has declined steadily by 78 percent but since 2015 it is largely flat. The latest figures show the typical turbine rates in China for about USD 500/kW, and elsewhere for USD 855/kW.

Figure 9 derived from IRENA statistics [152] gives information on employment numbers for renewable energies. Overall wind employment in the world is projected at 510,000 workers rapidly, with Germany (140,800 jobs) and the US rising 8%, to a new peak of 114,000 employees.

The European Union plan for a sustainable eco-friendly economy offers options for holding global warming below 2°C [153]. The pursuit of aggressive carbon goals requires widespread use of renewable sources of energy. The research shows that a mandatory one-third of CO_2 emissions are due to power generation and an increase in the number of wind power plants is expected to reach 12% of global electricity production within twenty years [154]. The research reports show that renewable energy sources are considered obligatory. China, the world's greatest user of fossil fuels and greenhouse gas generator, faces enormous social and economic changes with unparalleled challenges.

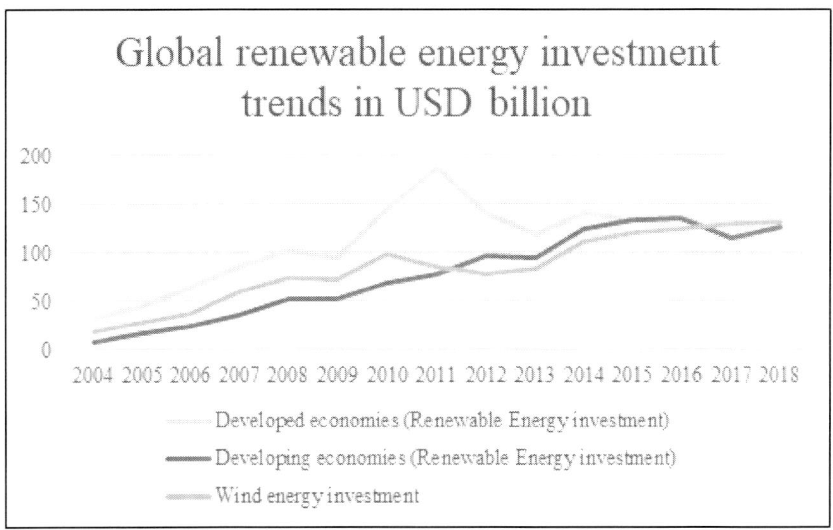

Figure 7. Global renewable energy investment trends.

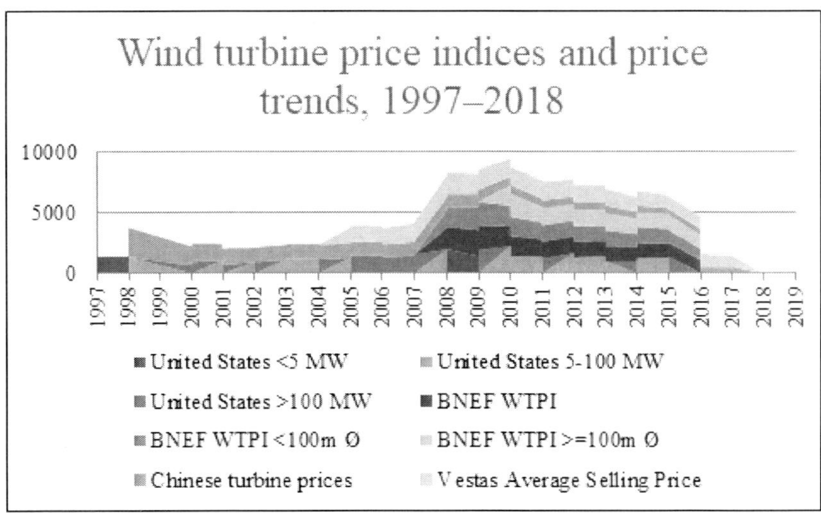

Figure 8. Wind turbine price indices and price trends, 1997–2018.

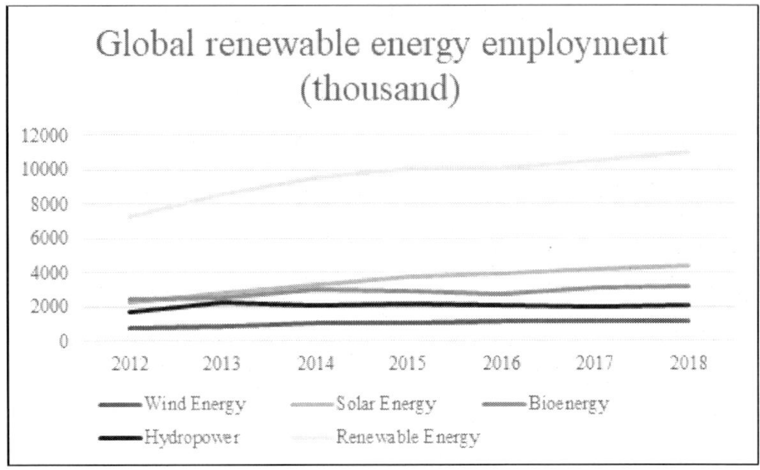

Figure 9. Total employment figures in the renewable energy sector by technology.

The cost of wind energy compared to fossil fuels costs is particularly controversial. The financial stimulus provided for both power-based wind farms and small residential wind turbines. With increasing demands for electricity along with a strong need to generate more energy from renewable sources; governments liberalized the market in order to invite more players to enter this sector by regulation the marketing regime [154]. Moreover, support policies of renewable energy integrated to buy the generated electricity according to the "feed-in tariffs". Additionally economic consequences of the turbine phenomenon and improved economic performance are evident [155]. To date, renewable energies have been of rather marginal relevance to the energy sector. Consequently, EU regulations governing the total share of renewable energies in the overall volume of energy produced and consumed are far from being attained. It is with this in mind that the energy market has a major interest in creating a new legal environment which boosts the market while allowing the state to comply with its obligations towards climate concern as fast as possible in a manner that can be financed.

Another important issue is the integration of wind energy. Since the cost effectiveness and strength against fossil fuel is relatively slow, it needs a massive support from public institutions. The inexhaustible, nonpolluting, somehow costless wind energy which does not require external procurement is the main new, renewable and alternate energy resource. Today where the electric energy consumption per person indicate development, continuous exhaustion of primary energy resources, ecological deterioration and even climatic balance concerns associated with burning of those fuels appear as a problem that needs an immediate solution. One of the solutions for the problem is to find clean and new energy resources, and second solution is to reduce consumption to the minimum economically acceptable level. Increasing demand for environmental products and services in energy market particularly for waste management, water supply and management and air pollution control sub-fields as consequence of economic growth, industrialization and urbanization.

Wind energy reliability is under improvement which affects the distribution system directly causing an electricity cost fluctuation. World electricity markets are constituted of principle elements such as production, transmission, distribution and supply. Because electrical energy cannot be stored like other products and has to be consumed as soon as it is generated, correlation of those elements should not be ignored when establishing and operating energy markets. Electric market models that users can benefit in the most efficient manner offering opportunities for participants is a view adopted in energy business. Individual structuring and privatization efforts of production, transmission, distribution, trade, etc. monopoly companies are accelerated in the World and particularly in European energy markets which are liberalizing and restructuring in this direction in order to create reliable and more functional electric markets. However, due to high capital requirement of transmission investments and significant expropriation difficulties, supply of transmission services mostly by public and monopoly tendency somewhat continues. The expectations regarding an increase in electricity demand and gross domestic product (GDP) per capita by increasing population and urbanization by acceleration of economic growth after the global economic crisis of which impacts are diminishing in time depicts a large potential in terms of electric energy demand. In other words, electric energy demand is anticipated to increase in folds in the medium and long term after crisis, and consumption values per person is expected to reach higher values by increase in GDP per capita.

The integration of massive renewable energy volumes into the power grid has caused many technological and economic obstacles for grid operators, utilities and market participants. The development of the industry in the coming years is expected to create major growth opportunities for Europe, if long-term visibility and stable regulatory frameworks are in place. The cost of offshore technologies also needs to become more competitive – and considerable investment in research and development into turbines, supply chain optimization, transmission, operations and maintenance is likely to be needed to help achieve cost reduction targets. Many of the solutions lie in the industry's own hands, and the industry is already working hard to make progress in cost reduction. The pace of growth in the industry now needs to be matched by an equal pace in lowering costs.

For a more functional electric market, the current official policy is towards gradual parting of public from investor in electric industry, except for transmission and privatization of the facilities in its possession, strengthening regulator position of public by having private entrepreneurs make the necessary investments in a competitive market environment and ensuring supply reliability. Various factors affect how and where jobs are created in the supply chain of renewable energy. These include government policies, supply-chain diversification, trading patterns and developments in reorganizing business and consolidating it. Apart from these reasons, the efficiency of labor is increasing over time. With the maturing of renewables sectors, growing economies of scale, overcoming learning curves and shifting to automated processes, less people are required for a particular task. Electric generation using wind power together with other renewables is important in order for countries to be challenging in advanced energy technologies field, to increase life quality by reducing environmental pollution, conformity process of energy industry to environmental statutes and to reduce dependency on external resources.

CONCLUSION

There are several parameters inflicting both the current status and future projections of wind energy which are listed in the logical framework matrix on risks and opportunities given in Table 2 in relation with the aspects investigated in this chapter.

Due to increasing environmental awareness, renewable energy technologies are rapidly improving. Ecological and environmental concerns which should be assessed in connection with social and economic impacts led to policy changes, which consequently triggered technological advancements in renewable energies, including wind energy, in an attempt to prevent impacts resulting from fossil fuel consumption. Environmental regulation and laws adopted by governments are shaped in accordance with social and environmental implications, targeting to keep emission levels under agreed protocols while supporting institutional environment for transition to a low carbon economy. This mechanism created new job opportunities and affected the overall economic growth.

Research and development activities focusing on renewable energies recently focuses on the integration of renewable resources to the energy mix, supported with smart grid technologies, software tools, design and management models. Increased share of renewables is considered a positive step towards decarbonization which in parallel is expected to result in improved energy access and security and ecosystems in general.

Especially in undeveloped countries access to electricity is considered a weakness, however funding support, feed-in tariffs and tax policies are applied for decreasing the high initial cost of energy technologies, promoting renewable energy technologies. Lack of potential to compete with conventional energy sources could be balanced with political stability combined with the resolution of social disputes accelerating technical maturity. Despite structural weaknesses and potential threats associated with every aspect of wind energy, valid strengths and promising opportunities contribute to its development.

Table 2. Logical framework matrix according to determined aspects of wind energy

		Social	Technical	Economic	Environmental	Political
Strengths	Top 5	Health consciousness	R&D activity	New jobs	Global awareness	Environmental regulations
		Emphasis on safety	Rate of technological change	Carbon economy	Decarbonization	Employment laws
		Raising Awareness	Rapid diversification	Economic growth	Food and water security	Law support by government
		Well educated population	Combined with other RES technologies	Funding support	Improvement in habitat	
		Energy democracy	Small scale wind power plants		Reduction of GHG emissions	
	Others	equality expectation, sustainable environment	machine learning		carbon sequestration, offset carbon footprint	regulatory standards, supportive institutional environment for low carbon economy
Weaknesses	Top 5	Population growth rate	Technology incentives	Economic growth	Ecological collapse	Tax policy
		Age distribution	Energy security	Exchange rates	Geographical limitation	Political instability
		Psychological impact	Power grid connection	Stagflation	Landscape usage	Trade restrictions and tariffs
		Aesthetic distribution	Conventional energy technologies	High battery costs	Chemical emissions from offshore wind farms	Systems for providing low-carbon energy
		Conflicts with local communities	Intermittent nature	Expensive wind metering	High noise levels	Lack of policy
	Others	shadow flicker, voice disturbances, visual impacts	high storage requirements, frequency and voltage fluctuations, design methods	market and socio-political factors, trade technologies taxes, unfavorable electricity prices	effect on wildlife, marine life and local bird abundance	

		Social	Technical	Economic	Environmental	Political
Opportunities	Top 5	Career attitudes	Technology advancement	Interest rates	Holding emissions under protocol levels	Carbon tax
		Easy access to electricity	Energy demand	Inflation rate	Wind versus fossil fuels	Energy and climate change priority
		Improvement of undeveloped or developing countries	Smart grid	Low battery costs in future	Carbon market	Emission trading system
		Public participation programs	Software tools and design models	Well-being	Protection of biodiversity	FITs
		Financial participation to community	Hybrid systems	Financial capital	Carbon footprint	Carbon market
	Others		automation, embodied AI, development of floating windmills, adaptability of modular wind turbines, grid-scale energy storage, power quality	energy consumption	environmental monitoring, "low-noise" foundation concept for OWFs, development of carbon price forecasts, decreasing CO2 and global temperature	tradable certificate systems
Threats	Top 5	Lack of education	Solar PV	Maturity of technology	Carbon price	Conflicts/War
		Potential of jobs lost and well-fare	High investment costs	Poverty	Ocean conservation	Fossil and nuclear lobby actions
		Social acceptance	Difficulties in grid connections	High capital investment	Characterization of wind velocity	LCA of renewables
		NIMBY	Competition with other RES	High production cost	GHG emissions in production	FITs
		High electricity prices	Competition with fossil fuels	Economic collapse	Climate change	Financial pressure imposed by subsidies
	Others	procedural, distributional and market acceptance, lack of information on technologies, mistrust in investors		unemployment	rising sea levels, shrinking Artic sea ice, deforestation, smog pollution, changing wind regime	

In view of the logical framework for risks and opportunities, the interlocking characteristics of social, political, economic, technical and environmental aspects suggest that a holistic approach is required for assessing the current status and future projections. It is therefore crucial to acknowledge the significance of each element and their intertwining structure in order to accurately monitor the development of wind energy.

REFERENCES

[1]	Igliński, Bartłomiej, Anna Iglińska, Grzegorz Koziński, Mateusz Skrzatek, Roman Buczkowski. "Wind Energy in Poland - History, Current State, Surveys, Renewable Energy Sources Act, SWOT Analysis." *Renewable and Sustainable Energy Reviews*, 2016. https://doi.org/ 10.1016/j.rser.2016.05.081.

[2]	Kaldellis, John K; Zafirakis, D. "The Wind Energy (r)Evolution: A Short Review of a Long History." *Renewable Energy*, 2011. https://doi.org/ 10.1016/j.renene.2011.01.002.

[3]	Leung, Dennis YC; Yuan Yang. "Wind Energy Development and Its Environmental Impact: A Review." *Renewable and Sustainable Energy Reviews*, 2012. https://doi.org/10.1016/j.rser.2011.09.024.

[4]	Ackermann, Thomas, Lennart Söder. "An Overview of Wind Energy-Status 2002." *Renewable and Sustainable Energy Reviews*, 6, no. 1–2, (January 1, 2002), 67–127. https://doi.org/10.1016/S1364-0321(02)00008-4.

[5]	REN 21 Renewables Now. *Renewables Global Status Report 2019. Galvanotechnik*, 2019.

[6]	IRENA. *Renewable Energy Statistics 2019. International Renewable Energy Agency*, 2019.

[7]	Linnenluecke, Martina K; Jianlei Han, Zheyao Pan, Tom Smith. "How Markets Will Drive the Transition to a Low Carbon Economy." *Economic Modelling*, 2019. https://doi.org/10.1016/ j.econmod.2018.07.010.

[8] Shimada, Koji, Yoshitaka Tanaka, Kei Gomi, Yuzuru Matsuoka. "Developing a Long-Term Local Society Design Methodology towards a Low-Carbon Economy: An Application to Shiga Prefecture in Japan." *Energy Policy*, 35, no. 9, (September 1, 2007), 4688–4703. https://doi.org/10.1016/J.ENPOL.2007.03.025.

[9] Przychodzen, Wojciech, Justyna Przychodzen. "Determinants of Renewable Energy Production in Transition Economies: A Panel Data Approach." *Energy*, 2020. https://doi.org/10.1016/j.energy.2019.116583.

[10] Lyu, Peng-hui, Eric WT. Ngai, Pei-yi Wu. "Scientific Data-Driven Evaluation on Academic Articles of Low-Carbon Economy." *Energy Policy*, 125, (February 1, 2019), 358–67. https://doi.org/10.1016/ J.ENPOL.2018.11.004.

[11] Daim, Tugrul, Ibrahim Iskin, Xin Li, Casey Zielsdorff, Ayse Elvan Bayraktaroglu, Turkay Dereli, Alptekin Durmusoglu. "Patent Analysis of Wind Energy Technology Using the Patent Alert System." *World Patent Information*, 2012. https://doi.org/ 10.1016/j.wpi.2011.11.001.

[12] Eck, Nees Jan van, Ludo Waltman. "Software Survey: VOSviewer, a Computer Program for Bibliometric Mapping." *Scientometrics*, 2010. https://doi.org/10.1007/s11192-009-0146-3.

[13] Lindman, Åsa, Patrik Söderholm. "Wind Energy and Green Economy in Europe: Measuring Policy-Induced Innovation Using Patent Data." *Applied Energy*, 2016. https://doi.org/10.1016/j.apenergy. 2015.10.128.

[14] Madvar, Mohammad Dehghani, Farzin Ahmadi, Reza Shirmohammadi, Alireza Aslani. "Forecasting of Wind Energy Technology Domains Based on the Technology Life Cycle Approach." *Energy Reports*, 2019. https://doi.org/10.1016/ j.egyr.2019.08.069.

[15] Georgilakis, Pavlos S. "Technical Challenges Associated with the Integration of Wind Power into Power Systems." *Renewable and Sustainable Energy Reviews*, 2008. https://doi.org/10.1016/ j.rser.2006.10.007.

[16] Liu, Tong, Gang Xu, Peng Cai, Longhu Tian, Qili Huang. "Development Forecast of Renewable Energy Power Generation in China and Its Influence on the GHG Control Strategy of the Country." *Renewable Energy*, 2011. https://doi.org/10.1016/ j.renene.2010.09.020.

[17] Schubel, Peter J; Richard J. Crossley. "Wind Turbine Blade Design." *Energies*, 2012. https://doi.org/10.3390/en5093425.

[18] Ancona, Dan, McVeigh, J. "Wind Turbine-Materials and Manufacturing Fact Sheet." *Princeton Energy Resources International, LLC*, 2001.

[19] Manganhar, Abdul Latif, Altaf Hussain Rajpar, Muhammad Ramzan Luhur, Saleem Raza Samo, Mehtab Manganhar. "Performance Analysis of a Savonius Vertical Axis Wind Turbine Integrated with Wind Accelerating and Guiding Rotor House." *Renewable Energy*, 136, (June 1, 2019), 512–20. https://doi.org/10.1016/ J.RENENE.2018.12.124.

[20] Barlas, TK; van Kuik, GAM. "Review of State of the Art in Smart Rotor Control Research for Wind Turbines." *Progress in Aerospace Sciences*, 2010. https://doi.org/10.1016/j.paerosci.2009.08.002.

[21] Polinder, Henk, Sjoerd WH. De Haan, Maxime R. Dubois, Johannes G. Slootweg. "Basic Operation Principles and Electrical Conversion Systems of Wind Turbines." *EPE Journal (European Power Electronics and Drives Journal)*, 2005. https://doi.org/10.1080/ 09398368.2005.11463604.

[22] Ibrahim, H; Ghandour, M; Dimitrova, M; Ilinca, A; Perron, J. "Integration of Wind Energy into Electricity Systems: Technical Challenges and Actual Solutions." In *Energy Procedia*, 2011. https://doi.org/ 10.1016/j.egypro.2011.05.092.

[23] Reder, MD; Gonzalez, E; Melero, JJ. "Wind Turbine Failures - Tackling Current Problems in Failure Data Analysis." In *Journal of Physics: Conference Series*, 2016. https://doi.org/10.1088/1742-6596/753/7/ 072027.

[24] Shafiullah, GM; Amanullah, MtOo; Shawkat Ali, ABM; Peter Wolfs. "Potential Challenges of Integrating Large-Scale Wind

Energy into the Power Grid-A Review." *Renewable and Sustainable Energy Reviews*, 2013. https://doi.org/10.1016/j.rser.2012.11.057.

[25] Haque, MH. "A Novel Method of Evaluating Performance Characteristics of a Self-Excited Induction Generator." *IEEE Transactions on Energy Conversion*, 2009. https://doi.org/10.1109/TEC.2009.2016124.

[26] Wang, Yuxuan, Tianye Sun. "Life Cycle Assessment of CO_2 Emissions from Wind Power Plants: Methodology and Case Studies." *Renewable Energy*, 2012. https://doi.org/10.1016/j.renene.2011.12.017.

[27] Singh, Bharat, Singh, SN. "Wind Power Interconnection into the Power System: A Review of Grid Code Requirements." *Electricity Journal*, 2009. https://doi.org/10.1016/j.tej.2009.04.008.

[28] Ayodele, TR; Jimoh, AA; Munda, JL; Agee, JT. "Challenges of Grid Integration of Wind Power on Power System Grid Integrity: A Review." *International Journal of Renewable Energy Research*, 2012. https://doi.org/10.20508/ijrer.87947.

[29] Denholm, Paul, Ramteen Sioshansi. "The Value of Compressed Air Energy Storage with Wind in Transmission-Constrained Electric Power Systems." *Energy Policy*, 2009. https://doi.org/10.1016/j.enpol.2009.04.002.

[30] Gutierrez, JJ; Ruiz, J; Saiz, P; Azcarate, I; Leturiondo, LA; Lazkano, A. *"Power Quality in Grid-Connected Wind Turbines,"* 2011.

[31] Shen, Xin, Xiaocheng Zhu, Zhaohui Du. "Wind Turbine Aerodynamics and Loads Control in Wind Shear Flow." *Energy*, 2011. https:// doi.org/10.1016/j.energy.2011.01.028.

[32] Wu, Jiasi, Buhan Zhang, Yazhou Jiang, Pei Bie, Hang Li. "Chance-Constrained Stochastic Congestion Management of Power Systems Considering Uncertainty of Wind Power and Demand Side Response." *International Journal of Electrical Power and Energy Systems*, 2019. https://doi.org/10.1016/j.ijepes.2018.12.026.

[33] Gao, Jianwei, Zeyang Ma, Fengjia Guo. "The Influence of Demand Response on Wind-Integrated Power System Considering

Participation of the Demand Side." *Energy*, 2019. https://doi.org/10.1016/ j.energy.2019.04.104.

[34] Alham, MH; Elshahed, M; Doaa Khalil Ibrahim, Essam El Din Abo El Zahab. "Optimal Operation of Power System Incorporating Wind Energy with Demand Side Management." *Ain Shams Engineering Journal*, 2017. https://doi.org/10.1016/j.asej.2015.07.004.

[35] Ni, Linna, Fushuan Wen, Weijia Liu, Jinling Meng, Guoying Lin, Sanlei Dang. "Congestion Management with Demand Response Considering Uncertainties of Distributed Generation Outputs and Market Prices." *Journal of Modern Power Systems and Clean Energy*, 2017. https://doi.org/10.1007/s40565-016-0257-9.

[36] Panwar, NL; Kaushik, SC; Surendra Kothari. "Role of Renewable Energy Sources in Environmental Protection: A Review." *Renewable and Sustainable Energy Reviews*, 2011. https://doi.org/10.1016/ j.rser.2010.11.037.

[37] Owusu, Phebe Asantewaa, Samuel Asumadu-Sarkodie. "A Review of Renewable Energy Sources, Sustainability Issues and Climate Change Mitigation." *Cogent Engineering*, 2016. https://doi.org/10.1080/ 23311916.2016.1167990.

[38] Ren, Guorui, Jinfu Liu, Jie Wan, Yufeng Guo, Daren Yu. "Overview of Wind Power Intermittency: Impacts, Measurements, and Mitigation Solutions." *Applied Energy*, 2017. https://doi.org/10.1016/ j.apenergy.2017.06.098.

[39] Marano, Vincenzo, Gianfranco Rizzo, Francesco Antonio Tiano. "Application of Dynamic Programming to the Optimal Management of a Hybrid Power Plant with Wind Turbines, Photovoltaic Panels and Compressed Air Energy Storage." *Applied Energy*, 2012. https://doi.org/10.1016/j.apenergy.2011.12.086.

[40] Sun, Hao, Xing Luo, Jihong Wang. "Feasibility Study of a Hybrid Wind Turbine System - Integration with Compressed Air Energy Storage." *Applied Energy*, 2015. https://doi.org/10.1016/ j.apenergy.2014.06.083.

[41] Foley, Aoife M; Paul G. Leahy, Antonino Marvuglia, Eamon J. McKeogh. "Current Methods and Advances in Forecasting of Wind

Power Generation." *Renewable Energy*, 2012. https://doi.org/ 10.1016/j.renene.2011.05.033.

[42] Xu, Jiuping, Li Li, Bobo Zheng. "Wind Energy Generation Technological Paradigm Diffusion." *Renewable and Sustainable Energy Reviews*, 2016. https://doi.org/10.1016/j.rser.2015.12.271.

[43] Weinzettel, Jan, Marte Reenaas, Christian Solli, Edgar G. Hertwich. "Life Cycle Assessment of a Floating Offshore Wind Turbine." *Renewable Energy*, 2009. https://doi.org/10.1016/j.renene. 2008.04.004.

[44] Dimitrov, Nikolay, Anand Natarajan, Jakob Mann. "Effects of Normal and Extreme Turbulence Spectral Parameters on Wind Turbine Loads." *Renewable Energy*, 2017. https://doi.org/10.1016/ j.renene.2016.10.001.

[45] Emejeamara, FC; Tomlin, AS. "A Method for Estimating the Potential Power Available to Building Mounted Wind Turbines within Turbulent Urban Air Flows." *Renewable Energy*, February 7, 2020. https://doi.org/10.1016/J.RENENE.2020.01.123.

[46] Kuik, GAM. van, Peinke, J; Nijssen, R; Lekou, D; Mann, J; Sørensen, JN; Ferreira, C; et al. "Long-Term Research Challenges in Wind Energy – a Research Agenda by the European Academy of Wind Energy." *Wind Energy Science*, 2016. https://doi.org/10.5194/wes-1-1-2016.

[47] Lubitz, William David. "Impact of Ambient Turbulence on Performance of a Small Wind Turbine." *Renewable Energy*, 2014. https://doi.org/ 10.1016/j.renene.2012.08.015.

[48] Tchakoua, Pierre, René Wamkeue, Mohand Ouhrouche, Fouad Slaoui-Hasnaoui, Tommy Andy Tameghe, Gabriel Ekemb. "Wind Turbine Condition Monitoring: State-of-the-Art Review, New Trends, and Future Challenges." *Energies*, 2014. https://doi.org/10.3390/ en7042595.

[49] Sina Kuseyri, İbrahim. "Condition Monitoring of Wind Turbines: Challenges and Opportunities," n.d.

[50] Kumar, Yogesh, Jordan Ringenberg, Soma Shekara Depuru, Vijay K. Devabhaktuni, Jin Woo Lee, Efstratios Nikolaidis, Brett

Andersen, Abdollah Afjeh. "Wind Energy: Trends and Enabling Technologies." *Renewable and Sustainable Energy Reviews*, 2016. https://doi.org / 10.1016/j.rser.2015.07.200.

[51] Mishnaevsky, Leon. "Repair of Wind Turbine Blades: Review of Methods and Related Computational Mechanics Problems." *Renewable Energy*, 2019. https://doi.org/10.1016/j.renene.2019.03.113.

[52] Diógenes, Jamil Ramsi Farkat, João Claro, José Coelho Rodrigues, Manuel Valentim Loureiro. "Barriers to Onshore Wind Energy Implementation: A Systematic Review." *Energy Research and Social Science*, 60, no. October 2018, (2020), 101337. https://doi.org/ 10.1016/j.erss.2019.101337.

[53] Kavari, Ghazale, Mojtaba Tahani, Mojtaba Mirhosseini. "Wind Shear Effect on Aerodynamic Performance and Energy Production of Horizontal Axis Wind Turbines with Developing Blade Element Momentum Theory." *Journal of Cleaner Production*, 219, (May 10, 2019), 368–76. https://doi.org/10.1016/J.JCLEPRO.2019.02.073.

[54] Moriarty, Patrick J; Sandy B. Butterfield. "Wind Turbine Modeling Overview for Control Engineers." In *Proceedings of the American Control Conference*, 2009. https://doi.org/10.1109/ ACC.2009.5160521.

[55] Whale, J; McHenry, MP; Malla, A. "Scheduling and Conducting Power Performance Testing of a Small Wind Turbine." *Renewable Energy*, 2013. https://doi.org/10.1016/j.renene.2012.11.032.

[56] Pagnini, Luisa C; Massimiliano Burlando, Maria Pia Repetto. "Experimental Power Curve of Small-Size Wind Turbines in Turbulent Urban Environment." *Applied Energy*, 2015. https://doi.org/10.1016/ j.apenergy.2015.04.117.

[57] Tummala, Abhishiktha, Ratna Kishore Velamati, Dipankur Kumar Sinha, Indraja, V; Hari Krishna, V. "A Review on Small Scale Wind Turbines." *Renewable and Sustainable Energy Reviews*, 2016. https://doi.org/10.1016/j.rser.2015.12.027.

[58] AWEA Small wind turbine Comiittee. "The U.S. Small Wind Turbine Industry. ROADMAP." *Awea Small Wind Turbine Commitee*, 2002.

[59] Li, Hao, Qingsong An, Bo Yu, Jun Zhao, Ling Cheng, Yan Wang. "Strategy Analysis of Demand Side Management on Distributed Heating Driven by Wind Power." In *Energy Procedia*, 2017. https://doi.org/10.1016/j.egypro.2017.03.624.

[60] Xu, Fangqiu, Jicheng Liu, Shuaishuai Lin, Qiongjie Dai, Cunbin Li. "A Multi-Objective Optimization Model of Hybrid Energy Storage System for Non-Grid-Connected Wind Power: A Case Study in China." *Energy*, 2018. https://doi.org/10.1016/j.energy.2018.08.152.

[61] Kong, Junhyuk, Sung Tae Kim, Byung O. Kang, Jaesung Jung. "Determining the Size of Energy Storage System to Maximize the Economic Profit for Photovoltaic and Wind Turbine Generators in South Korea." *Renewable and Sustainable Energy Reviews*, 2019. https://doi.org/10.1016/j.rser.2019.109467.

[62] Liu, Ye, Xiaogang Wu, Jiuyu Du, Ziyou Song, Guoliang Wu. "Optimal Sizing of a Wind-Energy Storage System Considering Battery Life." *Renewable Energy*, 147, (March 1, 2020), 2470–83. https://doi.org/ 10.1016/J.RENENE.2019.09.123.

[63] Miao, Di, Sarmistha Hossain. "Improved Gray Wolf Optimization Algorithm for Solving Placement and Sizing of Electrical Energy Storage System in Micro-Grids." *ISA Transactions*, February 15, 2020. https://doi.org/10.1016/J.ISATRA.2020.02.016.

[64] Aslam, Sheraz, Adia Khalid, Nadeem Javaid. "Towards Efficient Energy Management in Smart Grids Considering Microgrids with Day-Ahead Energy Forecasting." *Electric Power Systems Research*, 2020. https://doi.org/10.1016/j.epsr.2020.106232.

[65] Sovacool, Benjamin K. "Rejecting Renewables: The Socio-Technical Impediments to Renewable Electricity in the United States." *Energy Policy*, 2009. https://doi.org/10.1016/j.enpol.2009.05.073.

[66]   Ellis, Geraint, Gianluca Ferraro. "The Social Acceptance of Wind Energy." *European Commission - JRC Science for Policy Report*, 2016. https://doi.org/10.2789/696070.

[67]   Wolsink, Maarten. "Wind Power Wind Power : Basic Challenge Concerning Social Acceptance Wind Power Social Acceptance." In *Encyclopedia of Sustainability Science and Technology*, 2012. https://doi.org/10.1007/978-1-4419-0851-3_88.

[68]   Bauwens, Thomas. "The Effect of Cooperative Ownership on Social Acceptance of Onshore Wind Power: A Multi-Method Analysis." *2Nd Interdisciplinary Symposium on Sustainable Development*, no. April, (2015), 1–31.

[69]   Romice, Ombretta, Kevin Thwaites, Sergio Porta, Mark Greaves. "Urban Design and Quality of Life Chapter 14 Urban Design and Quality of Life." *Sustainable Behavior and Quality of Life*, no. August, 2017 (2016), 241–73. https://doi.org/10.1007/978-3-319-31416-7.

[70]   Wüstenhagen, Rolf, Maarten Wolsink, Mary Jean Bürer. "Social Acceptance of Renewable Energy Innovation: An Introduction to the Concept." *Energy Policy*, 2007. https://doi.org/10.1016/j.enpol.2006.12.001.

[71]   Rosso-Cerón, Ana María, Viatcheslav Kafarov. "Barriers to Social Acceptance of Renewable Energy Systems in Colombia." *Current Opinion in Chemical Engineering*, 10, (November 1, 2015), 103–10. https://doi.org/10.1016/J.COCHE.2015.08.003.

[72]   Devine-Wright, Patrick, Bouke Wiersma. "Understanding Community Acceptance of a Potential Offshore Wind Energy Project in Different Locations: An Island-Based Analysis of 'Place-Technology Fit.'" *Energy Policy*, 2020. https://doi.org/10.1016/j.enpol.2019.111086.

[73]   Roddis, Philippa, Stephen Carver, Martin Dallimer, Paul Norman, Guy Ziv. "The Role of Community Acceptance in Planning Outcomes for Onshore Wind and Solar Farms: An Energy Justice Analysis." *Applied Energy*, 2018. https://doi.org/10.1016/j.apenergy.2018.05.087.

[74] Dimitropoulos, Alexandros, Andreas Kontoleon. "Assessing the Determinants of Local Acceptability of Wind-Farm Investment: A Choice Experiment in the Greek Aegean Islands." *Energy Policy*, 37, no. 5, (2009), 1842–54. https://doi.org/10.1016/j.enpol.2009.01.002.

[75] Vuichard, Pascal, Alexander Stauch, Nathalie Dällenbach. "Individual or Collective? Community Investment, Local Taxes, and the Social Acceptance of Wind Energy in Switzerland." *Energy Research & Social Science*, 2019. https://doi.org/10.1016/j.erss.2019.101275.

[76] Landeta-Manzano, Beñat, Germán Arana-Landín, Pilar M. Calvo, Iñaki Heras-Saizarbitoria. "Wind Energy and Local Communities: A Manufacturer's Efforts to Gain Acceptance." *Energy Policy*, 2018. https://doi.org/10.1016/j.enpol.2018.05.034.

[77] Khorsand, Iman, Christine Kormos, Erin G. Macdonald, Curran Crawford. "Wind Energy in the City: An Interurban Comparison of Social Acceptance of Wind Energy Projects." *Energy Research and Social Science*, 2015. https://doi.org/10.1016/j.erss.2015.04.008.

[78] Walker, Benjamin JA; Bouke Wiersma, Etienne Bailey. "Community Benefits, Framing and the Social Acceptance of Offshore Wind Farms: An Experimental Study in England." *Energy Research and Social Science*, 2014. https://doi.org/10.1016/j.erss.2014.07.003.

[79] Sposato, Robert Gennaro, Nina Hampl. "Worldviews as Predictors of Wind and Solar Energy Support in Austria: Bridging Social Acceptance and Risk Perception Research." *Energy Research and Social Science*, 2018. https://doi.org/10.1016/j.erss.2018.03.012.

[80] Wilson, Geoff A; Sarah L. Dyke. "Pre- and Post-Installation Community Perceptions of Wind Farm Projects: The Case of Roskrow Barton (Cornwall, UK)." *Land Use Policy*, 2016. https://doi.org/10.1016/ j.landusepol.2015.12.008.

[81] Motosu, Memi, Yasushi Maruyama. "Local Acceptance by People with Unvoiced Opinions Living Close to a Wind Farm: A Case

Study from Japan." *Energy Policy*, 2016. https://doi.org/10.1016/ j.enpol.2016.01.018.

[82] Abolhosseini, Shahrouz, Almas Heshmati. "The Main Support Mechanisms to Finance Renewable Energy Development." *Renewable and Sustainable Energy Reviews*, 2014, https://doi.org/10.1016/ j.rser.2014.08.013.

[83] Tamás, Mészáros Mátyás, Bade Shrestha, SO; Huizhong Zhou. "Feed-in Tariff and Tradable Green Certificate in Oligopoly." *Energy Policy*, 2010. https://doi.org/10.1016/j.enpol.2010.03.028.

[84] Sun, Peng, Pu yan Nie. "A Comparative Study of Feed-in Tariff and Renewable Portfolio Standard Policy in Renewable Energy Industry." *Renewable Energy*, 2015. https://doi.org/10.1016/j.renene. 2014.08.027.

[85] Ringel, Marc. "Fostering the Use of Renewable Energies in the European Union: The Race between Feed-in Tariffs and Green Certificates." *Renewable Energy*, 31, no. 1, (January 1, 2006), 1–17. https://doi.org/ 10.1016/J.RENENE.2005.03.015.

[86] Langer, Katharina, Thomas Decker, Jutta Roosen, Klaus Menrad. "Factors Influencing Citizens' Acceptance and Non-Acceptance of Wind Energy in Germany." *Journal of Cleaner Production*, 2018. https://doi.org/10.1016/j.jclepro.2017.11.221.

[87] Huber, Stefanie; Horbaty, Robert. *"Technical Report Results of IEA Wind Task 28 on Social Acceptance of Wind Energy,"* 2010, 1–91.

[88] Kaltenegger, Oliver, Andreas Löschel, Martin Baikowski, Jörg Lingens. "Energy Costs in Germany and Europe: An Assessment Based on a (Total Real Unit) Energy Cost Accounting Framework." *Energy Policy*, 104, (May 1, 2017), 419–30. https://doi.org/10.1016/ J.ENPOL.2016.11.039.

[89] Hitaj, Claudia, Andreas Löschel. "The Impact of a Feed-in Tariff on Wind Power Development in Germany." *Resource and Energy Economics*, 2019. https://doi.org/10.1016/j.reseneeco.2018.12.001.

[90] Tanaka, Makoto, Yihsu Chen. "Market Power in Renewable Portfolio Standards." *Energy Economics*, 39, (September 1, 2013), 187–96. https://doi.org/10.1016/J.ENECO.2013.05.004.

[91] Pineda, Salvador, Andreas Bock. "Renewable-Based Generation Expansion under a Green Certificate Market." *Renewable Energy*, 91, (June 1, 2016), 53–63. https://doi.org/10.1016/ J.RENENE. 2015.12.061.

[92] Atănăsoae, Pavel, Radu Pentiuc. "Considerations on the Green Certificate Support Dystem for Electricity Production from Renewable Energy Sources." *Procedia Engineering*, 181, (January 1, 2017), 796–803. https://doi.org/10.1016/J.PROENG.2017.02.469.

[93] Kitzing, Lena, Nina Juul, Michael Drud, Trine Krogh Boomsma. "A Real Options Approach to Analyse Wind Energy Investments under Different Support Schemes." *Applied Energy*, 188, (February 15, 2017), 83–96. https://doi.org/10.1016/J.APENERGY.2016.11.104.

[94] Schusser, Sandra, Jūratė Jaraitė. "Explaining the Interplay of Three Markets: Green Certificates, Carbon Emissions and Electricity." *Energy Economics*, 2018. https://doi.org/10.1016/j.eneco. 2018.01.012.

[95] Hulshof, Daan, Catrinus Jepma, Machiel Mulder. "Performance of Markets for European Renewable Energy Certificates." *Energy Policy*, 128, (May 1, 2019), 697–710. https://doi.org/ 10.1016/J.ENPOL. 2019.01.051.

[96] Besseau, Romain, Romain Sacchi, Isabelle Blanc, Paula Pérez-López. "Past, Present and Future Environmental Footprint of the Danish Wind Turbine Fleet with LCA_WIND_DK, an Online Interactive Platform." *Renewable and Sustainable Energy Reviews*, 2019. https://doi.org/ 10.1016/j.rser.2019.03.030.

[97] Morini, Antonio Augusto, Manuel J. Ribeiro, Dachamir Hotza. "Early-Stage Materials Selection Based on Embodied Energy and Carbon Footprint." *Materials and Design*, 2019. https://doi.org/10.1016/ j.matdes.2019.107861.

[98] Kaldellis, JK; Apostolou, D. "Life Cycle Energy and Carbon Footprint of Offshore Wind Energy. Comparison with Onshore Counterpart." *Renewable Energy*, 108, (2017), 72–84. https://doi.org/10.1016/ j.renene.2017.02.039.

[99] Aso, Raymond, Wai Ming Cheung. "Towards Greener Horizontal-Axis Wind Turbines: Analysis of Carbon Emissions, Energy and Costs at the Early Design Stage." *Journal of Cleaner Production*, 2015. https://doi.org/10.1016/j.jclepro.2014.10.020.

[100] Demir, Nesrin, Akif Taşkin. "Life Cycle Assessment of Wind Turbines in Pınarbaşı-Kayseri." *Journal of Cleaner Production*, 54, (2013), 253–63. https://doi.org/10.1016/j.jclepro.2013.04.016.

[101] Uddin, Md. Shazib, Kumar, S. "Energy, Emissions and Environmental Impact Analysis of Wind Turbine Using Life Cycle Assessment Technique." *Journal of Cleaner Production*, 69, (April 15, 2014), 153–64. https://doi.org/10.1016/J.JCLEPRO.2014.01.073.

[102] Wang, Shifeng, Sicong Wang, Jinxiang Liu. "Life-Cycle Green-House Gas Emissions of Onshore and Offshore Wind Turbines." *Journal of Cleaner Production*, 2019. https://doi.org/10.1016/j.jclepro.2018.11.031.

[103] Intrinsik Corp. *"Greenhouse Gas Emissions Associated With Various Methods Of Power Generation In Ontario,"* no. October, (2016), 47. https://www.opg.com/darlington-refurbishment/
Documents/IntrinsikReport_GHG_OntarioPower.pdf.

[104] Zhang, Xiaoqin, Ling Xu, Yu Chen, Tingting Liu. "Emergy-Based Ecological Footprint Analysis of a Wind Farm in China." *Ecological Indicators*, 2020. https://doi.org/10.1016/j.ecolind.2019.106018.

[105] Hao, Yan, Chengshi Tian, Chunying Wu. "Modelling of Carbon Price in Two Real Carbon Trading Markets." *Journal of Cleaner Production*, 2020. https://doi.org/10.1016/j.jclepro.2019.118556.

[106] Zhao, Xiaoli, Jin Yao, Chuyu Sun, Wengeng Pan. "Impacts of Carbon Tax and Tradable Permits on Wind Power Investment in China." *Renewable Energy*, 2019. https://doi.org/10.1016/ j.renene. 2018.09.068.

[107] Jevnaker, Torbjørg, Jørgen Wettestad. "Ratcheting up Carbon Trade: The Politics of Reforming EU Emissions Trading."

Global Environmental Politics, 2017. https://doi.org/10.1162/ GLEP_a_00403.

[108] Lin, Boqiang, Zhijie Jia. "Energy, Economic and Environmental Impact of Government Fines in China's Carbon Trading Scheme." *Science of the Total Environment*, 2019. https://doi.org/10.1016/ j.scitotenv.2019.02.405.

[109] Hu, Yucai, Shenggang Ren, Yangjie Wang, Xiaohong Chen. "Can Carbon Emission Trading Scheme Achieve Energy Conservation and Emission Reduction? Evidence from the Industrial Sector in China." *Energy Economics*, 2020. https://doi.org/10.1016/j. eneco.2019.104590.

[110] Adedipe, Oyewole, Feargal Brennan, Athanasios Kolios. "Review of Corrosion Fatigue in Offshore Structures: Present Status and Challenges in the Offshore Wind Sector." *Renewable and Sustainable Energy Reviews*, 2016. https://doi.org/10.1016/j.rser.2016.02.017.

[111] Kim, Hyo Jin, Ju Hee Kim, Seung Hoon Yoo. "Social Acceptance of Offshore Wind Energy Development in South Korea: Results from a Choice Experiment Survey." *Renewable and Sustainable Energy Reviews*, 2019. https://doi.org/10.1016/j.rser.2019.109253.

[112] Failla, Giuseppe, Felice Arena. "New Perspectives in Offshore Wind Energy." *Philosophical Transactions of the Royal Society A: Mathematical, Physical and Engineering Sciences*, 373, no. 2035, (2015). https://doi.org/10.1098/rsta.2014.0228.

[113] Kirchgeorg, T; Weinberg, I; Hörnig, M; Baier, R; Schmid, MJ; Brockmeyer, B. "Emissions from Corrosion Protection Systems of Offshore Wind Farms: Evaluation of the Potential Impact on the Marine Environment." *Marine Pollution Bulletin*, 2018. https:// doi.org/10.1016/j.marpolbul.2018.08.058.

[114] Igwemezie, Victor, Ali Mehmanparast, Athanasios Kolios. "Current Trend in Offshore Wind Energy Sector and Material Requirements for Fatigue Resistance Improvement in Large Wind Turbine Support Structures – A Review." *Renewable and Sustainable Energy Reviews*, 2019. https://doi.org/10.1016/j.rser.2018.11.002.

[115] Loss, Scott R; Tom Will, Peter P. Marra. "Estimates of Bird Collision Mortality at Wind Facilities in the Contiguous United States." *Biological Conservation*, 168, (2013), 201–9. https://doi.org/10.1016/ j.biocon.2013.10.007.

[116] Salvador, Santiago, Luis Gimeno, Javier Sanz Larruga, F. "The Influence of Regulatory Framework on Environmental Impact Assessment in the Development of Offshore Wind Farms in Spain: Issues, Challenges and Solutions." *Ocean and Coastal Management*, 2018. https:// doi.org/10.1016/j.ocecoaman.2018.05.010.

[117] Kaldellis, JK; Apostolou, D; Kapsali, M; Kondili, E. "Environmental and Social Footprint of Offshore Wind Energy. Comparison with Onshore Counterpart." *Renewable Energy*, 92, (July 1, 2016), 543–56. https://doi.org/10.1016/J.RENENE.2016.02.018.

[118] Snyder, Brian, Mark J. Kaiser. "Ecological and Economic Cost-Benefit Analysis of Offshore Wind Energy." *Renewable Energy*, 2009. https://doi.org/10.1016/j.renene.2008.11.015.

[119] Stewart, Gavin B; Andrew S. Pullin, Christopher F. Coles. "Poor Evidence-Base for Assessment of Windfarm Impacts on Birds." *Environmental Conservation*, 2007. https://doi.org/10.1017/ S0376892907003554.

[120] Wu, Yunna, Yong Hu, Xiaoshan Lin, Lingwenying Li, Yiming Ke. "Identifying and Analyzing Barriers to Offshore Wind Power Development in China Using the Grey Decision-Making Trial and Evaluation Laboratory Approach." *Journal of Cleaner Production*, 189, (July 10, 2018), 853–63. https://doi.org/10.1016/ J.JCLEPRO.2018.04.002.

[121] Gasparatos, Alexandros, Christopher NH. Doll, Miguel Esteban, Abubakari Ahmed, Tabitha A. Olang. "Renewable Energy and Biodiversity: Implications for Transitioning to a Green Economy." *Renewable and Sustainable Energy Reviews*, 2017. https://doi.org/ 10.1016/j.rser.2016.08.030.

[122] Momber, A. "Corrosion and Corrosion Protection of Support Structures for Offshore Wind Energy Devices (OWEA)." *Materials*

and Corrosion 62, no. 5, (2011), 391–404. https://doi.org/10.1002/maco.201005691.

[123] Lyon, SB; Bingham, R; Mills, DJ. "Advances in Corrosion Protection by Organic Coatings: What We Know and What We Would like to Know." *Progress in Organic Coatings*, 102, (January 1, 2017), 2–7. https://doi.org/10.1016/J.PORGCOAT.2016.04.030.

[124] Price, Seth J; Rita B. Figueira. "Corrosion Protection Systems and Fatigue Corrosion in Offshore Wind Structures: Current Status and Future Perspectives." *Coatings*, 2017. https://doi.org/10.3390/coatings7020025.

[125] Sklenicka, Petr, Jan Zouhar. "Predicting the Visual Impact of Onshore Wind Farms via Landscape Indices: A Method for Objectivizing Planning and Decision Processes." *Applied Energy*, 2018. https:// doi.org/10.1016/j.apenergy.2017.11.027.

[126] Thackara, John. *In the Bubble: Designing in a Complex World. Choice Reviews Online.*, Vol. *43*, 2005. https://doi.org/10.5860/choice.43-1350.

[127] Koschinski, Sven, Karin Ludeman. "Development of Noise Mitigation Measures in Offshore Wind Farm Construction," no. February (2011). https://www.researchgate.net/publication/308110557_Development_of_Noise_Mitigation_Measures_in_Offshore_Wind_Farm_Construction.

[128] Thompson, Paul M; Gordon D. Hastie, Jeremy Nedwell, Richard Barham, Kate L. Brookes, Line S. Cordes, Helen Bailey, Nancy McLean. "Framework for Assessing Impacts of Pile-Driving Noise from Offshore Wind Farm Construction on a Harbour Seal Population." *Environmental Impact Assessment Review*, 2013. https://doi.org/ 10.1016/j.eiar.2013.06.005.

[129] Madsen, PT; Wahlberg, M; Tougaard, J; Lucke, K; Tyack, P. "Wind Turbine Underwater Noise and Marine Mammals: Implications of Current Knowledge and Data Needs." *Marine Ecology Progress Series*, 2006. https://doi.org/10.3354/meps309279.

[130] Saidur, R; Rahim, NA; Islam, MR; Solangi, KH. "Environmental Impact of Wind Energy." *Renewable and Sustainable Energy Reviews*, 2011. https://doi.org/10.1016/j.rser.2011.02.024.

[131] Kyoto Mechanisms. "Chapter 2 Design and Implementation of Market-Based Climate Policy." *Developments in Environmental Economics*, 7, no. C, (2004), 27–53. https://doi.org/10.1016/S0927-5207(04)80004-9.

[132] McDonagh, Shane, David M. Wall, Paul Deane, Jerry D. Murphy. "The Effect of Electricity Markets, and Renewable Electricity Penetration, on the Levelised Cost of Energy of an Advanced Electro-Fuel System Incorporating Carbon Capture and Utilisation." *Renewable Energy*, 2019. https://doi.org/10.1016/j.renene.2018.07.058.

[133] European Commission. "COM(2014) 15 Final: A Policy Framework for Climate and Energy in the Period from 2020 to 2030," no. 2014 (2012), 1–18.

[134] Quint, Dov, Steve Dahlke. "The Impact of Wind Generation on Wholesale Electricity Market Prices in the Midcontinent Independent System Operator Energy Market: An Empirical Investigation." *Energy*, 2019. https://doi.org/10.1016/j.energy.2018.12.028.

[135] IRENA International Renewable Energy Agency. *Renewable Power Generation Costs in 2017. International Renewable Energy Agency*, 2018. https://doi.org/10.1007/ SpringerReference_7300.

[136] Fan, Jing Li, Shijie Wei, Lin Yang, Hang Wang, Ping Zhong, Xian Zhang. "Comparison of the LCOE between Coal-Fired Power Plants with CCS and Main Low-Carbon Generation Technologies: Evidence from China." *Energy*, 2019. https://doi.org/10.1016/j.energy.2019.04.003.

[137] Fraunhofer, ISE. "*Fraunhofer ISE Annual Report 2011*," 2014.

[138] Peng, Donna, Rahmatallah Poudineh. "Electricity Market Design under Increasing Renewable Energy Penetration: Misalignments Observed in the European Union." *Utilities Policy*, 2019. https://doi.org/10.1016/ j.jup.2019.100970.

[139] Barbose, Galen, Ryan Wiser, Jenny Heeter, Trieu Mai, Lori Bird, Mark Bolinger, Alberta Carpenter; et al. "A Retrospective Analysis of Benefits and Impacts of U.S. Renewable Portfolio Standards." *Energy Policy*, 2016. https://doi.org/10.1016/j.enpol.2016.06.035.

[140] "The Global Wind Energy Council, 'Q1 2019 Global Wind Market in 2018,'" n.d.

[141] GWEC. "GWEC Global Wind 2017 Report - A Snapshot of Top Wind Markets in 2017: Offshore Wind," 2017. https://doi.org/10.1021/ jp0213102.

[142] Wind Europe. "Offshore Wind in Europe - Key Trends and Statistics 2017." *Refocus*, 2018. https://doi.org/10.1016/S1471-0846(02)80021-X.

[143] deCastro, M; Salvador, S; Gómez-Gesteira, M; Costoya, X; Carvalho, D; Sanz-Larruga, FJ; Gimeno, L. "Europe, China and the United States: Three Different Approaches to the Development of Offshore Wind Energy." *Renewable and Sustainable Energy Reviews*, 2019. https:// doi.org/10.1016/j.rser.2019.04.025.

[144] IRENA International Renewable Energy Agency. "Renewable Electricity Capacity and Generation Statistics," n.d.

[145] The International Energy Agency. "World Energy Outlook 2018," n.d.

[146] "Women_Empowerment_Sustainable_Energy_0," n.d.

[147] Kuik, Onno, Frédéric Branger, Philippe Quirion. "Competitive Advantage in the Renewable Energy Industry: Evidence from a Gravity Model." *Renewable Energy*, 2019. https://doi.org/10.1016/ j.renene.2018.07.046.

[148] Efstathiou, Jr. Jim. *"Global Clean Energy Funding Dips 8% as China Cools Solar Boom,"* 2019.

[149] IRENA International Renewable Energy Agency. "Investment Trends," n.d. http://resourceirena.irena.org/gateway/ dashboard/?topic=6&subTopic=11.

[150] Ajadi, Tayo, Rohan Boyle, David Strahan, Matthias Kimmel, Bryony Collins, Albert Cheung, Lisa Becker. "Global Trends in

Renewable Energy Investment 2019." *Bloomberg New Energy Finance*, 2019, 76.

[151] IRENA International Renewable Energy Agency. "Overview Tables," n.d. http://resourceirena.irena.org/gateway/ dashboard/?topic=6&subTopic=11.

[152] IRENA International Renewable Energy Agency. "Employment Time Series," n.d. http://resourceirena.irena.org/gateway/dashboard/?topic=7&subTopic=53.

[153] Gamel, Johannes, Klaus Menrad, Thomas Decker. "Is It Really All about the Return on Investment? Exploring Private Wind Energy Investors' Preferences." *Energy Research and Social Science*, 2016. https:// doi.org/10.1016/j.erss.2016.01.004.

[154] Lacal-Arántegui, Roberto. "Globalization in the Wind Energy Industry: Contribution and Economic Impact of European Companies." *Renewable Energy*, 134, (April 1, 2019), 612–28. https://doi.org/ 10.1016/j.renene.2018.10.087.

[155] Blanco, María Isabel. "The Economics of Wind Energy." *Renewable and Sustainable Energy Reviews*, 13, no. 6–7 (2009): 1372–82. https:// doi.org/10.1016/j.rser.2008.09.004.

BIOGRAPHICAL SKETCHES

Dilvin Çebi

Affiliation: Ege University Solar Energy Institute

Education: PhD Candidate, Ege University Solar Energy Institute

Business Address: Ege University Solar Energy Institute

Research and Professional Experience: Renewable energy technologies, energy efficiency, biofuels

Publications from the Last 3 Years:

Cebi, D., Sarptas, H. (2018) Recent Developments in Urban Energy Efficiency: A Review for Turkey, *GAPYENEV International GAP Renewable Energy and Energy Efficiency Congress*, *May 10 -12, 2018*, Şanlıurfa, Turkey.

Cebi, D., Sarptas, H. (2019) Evaluation of Landfill Gas Estimation Models with Respect to Actual LFG Production: A Case Study from Turkey, *X. International Multidisciplinary Congress of Eurasia, IMCOFE 2019, August 6-8, 2019,* Praque, Czechia.

Fikret Müge Alptekin

Affiliation: Ege University Solar Energy Institute

Education: PhD Student, Ege University Solar Energy Institute

Business Address: Ege University Solar Energy Institute

Research and Professional Experience: Renewable energy, biorefinery, biofuels, energy strorage devices

Merve Uyan

Affiliation: Ege University Solar Energy Institute

Education: PhD Student, Ege University Solar Energy Institute

Business Address: Ege University Solar Energy Institute

Research and Professional Experience: Renewable energy technologies, composite material, biofuels

Engin Deniz

Affiliation: The Beuth University of Applied Sciences Berlin

Education: The Beuth University of Applied Sciences Berlin, Master of Business Administration

Business Address: Hanwha Q Cells GmbH, Lorenzweg 5, 12099 Berlin, Germany

Research and Professional Experience: Renewable energy technologies

Melih Soner Çeliktaş, PhD

Affiliation: Ege University Solar Energy Institute

Education: PhD, Ege University Solar Energy Institute

Business Address: Ege University Solar Energy Institute

Research and Professional Experience: Melih Soner Celiktas is a mechanical Engineer who graduated from Yildiz Technical University. He obtained his Diploma (MSc) in Mechanical Engineering and his Doctorate (PhD) in Solar Energy Institute, in 2005 and 2009, respectively, both from the Ege University, Turkey. He is working as an associate professor at Ege University Solar Energy Institute. His research interests lies in the field of renewable energy and impacts upon the society and economy. He is currently investigating the biorefinery systems such as Physico-chemical pretreatment, bioconversion of renewable lignocellulosic biomass to biofuel and value added products like biomaterials using the latest conversion techniques and analysis.

Since last decade, he has been working (co-ordinated & collaborated) in a number of EU projects including IRC, FASCAP, EBIC, Posmetrans

SPINE and IQVETRES. His research interests include renewable energy, technology foresight, bibliometric analysis, biobased materials, energy economy, renewable energy policy and their implementation in technological and socio-economic fields. His future research plans are to build on the robotics and automation systems engineering on energy crops using biorefinery technics to further develop models and tools in conjunction with other related researchers and industrial actors.

Celiktas has published over 20 research papers in international journals. He has presented more than 60 papers in national and international meetings and scientific conferences. Celiktas is member of editorial board of scientific Journals and co-author of five books, more than 200 peer-review scientific papers published in international journals and more than 10 book chapters. He also has a national patent on biorefinery technology.

Publications from the Last 3 Years:

Gunes, K., Yaglikci, M., Celiktas, M.S. (2020). "Pressurized liquid hot water pretreatment and enzymatic hydrolysis of switchgrass aiming at the enhancement of fermentable sugars" *Biofuels,* DOI: 10.1080/17597269.2019.1706273, Accepted.

Celiktas, M. S., Alptekin F. M. (2019). "Conversion of model biomass to carbon-based material with high conductivity by using carbonization" *Energy*, 116089.

Celiktas, M. S., Yaglikci, M., Khosravi Maleki, F. (2019). "Subcritical water extraction derived lignin for creation of sustainable reinforced composite materials" *Polymer Testing*, 77.

Uyan, M., Alptekin, F .M., Bastabak, B., Ozgul, S. Erdogan, B., Ogut, T., Sezer, U. Celiktas, M. S. (2019). "Combined biofuel production from cotton stalk and seed with a biorefinery approach" *Biomass Conversion and Biorefinery*, Doi: 10.1007/s13399-019-00427-z.

Uyan, M., Celiktas, M. S. (2019). "Plant Fiber Reinforced Biocomposite: Properties And Applications" Mugla Journal of Science and Technology, 5(2), 42-48.

Cebi, D., Alptekin, F. M., Uyan, M., Sarptas, H., Celiktas, M. S. (2019). Bioethanol Production From Hazelnut Shell By Liquid Hot Water Pretreatment. *X. International Multidisciplinary Congress of Eurasia.*

Cebi, D., Alptekin, F. M., Uyan, M., Sarptas, H., Celiktas, M. S. (2019). "A Review on the Entrepreneurial Structure of Higher Education Institutions Regarding Academic Studies on Energy" *IITEEC 2019 International Instructional Technologies in Engineer ing Education Conference*, 1(1), 8-8.

Celiktas, M. S. (2019). "Value added products from bio-based manufacturing and the Future trends of bioconversion and biorefinery Industries" *MAS European International Congress on Mathematics-Engineering-Natural Medical Sciences-III*, 1(1), 255-261.

Celiktas, M. S. (2019). "Future energy systems based on smart energy infrastructure integrating renewable energy" *MAS European International Congress on Mathematics-Engineering-Natural Medical Sciences-III*, 1(1), 262-267.

Gultekin, S. Y., Olgun, H., Celiktas, M. S. (2018). "Comparison of solid biofuels produced from olive pomace with two different conversion methods: torrefaction and hydrothermal carbonization" *International Journal of Engineering Technology*, 7(2.23), 143, Doi: 10.14419/ijet.v7i2.23.11903

Ucar, R. C., Sengul, A., Celiktas, M. S. (2018). *Sustainable Recovery and Reutilization of Cereal Processing By-Products*, Chapter 4. "Wheat bran-based biorefinery" (2018)., Elsevier, Editor: Charis Galanakis, ISBN:9780081021620.

Alptekin, F. M., Cakır, M., Celiktas, M. S. (2018). "Supercapacitor As An Energy Storage Device: Current And Future Prospect" *SOLARTR*, 1, 45-58.

Uyan, M., Cakır M., Celiktas, M. S. (2018). "Plant fibre based biocomposites and the future course available." *SOLARTR*, 289-302.

Uyan, M., Celiktas, M. S. (2018). "Future prospects of biocomposites" IV. International Ege Composite Materials Symposium, *KOMPEGE* 2018, 1(1), 390-400.

Alptekin, F. M., Celiktas, M. S. (2018). "Nanocomposites and its applications in energyconversion and storage devices" IV. International Ege Composite Materials Symposium, *KOMPEGE* 2018, 1(1), 531-545.

Celiktas, M. S., Uyan, M., Alptekin, F. M. (2017). "Biorefinery concept: Current status and future prospects" *International Conference on Engineering Technologies (ICENTE'17)*, p27-32, Konya,Turkey.

Cerone, N., Zimbardi, F., Celiktas, M. S., Vito, V. (2017). "Pilot plant air steam gasification of nut shells for syngas production" presented at the *25th European Biomass Conference and Exhibition*, pages 843-846.

Deniz, E., Celiktas, M. S. (2017). "Overview of future for offshore wind energy in Turkey" *7th International 100 Renewable Energy Conference, IRENEC* 2017, p105-110.

In: Wind Speed: An Overview
Editor: Nicolas Kočí

ISBN: 978-1-53618-412-9
© 2020 Nova Science Publishers, Inc.

Chapter 2

AN ASSESSMENT OF RECENT STUDIES ON WIND SPEED FORECASTING METHODS FOR NATURAL HAZARDS IN DIFFERENT CLIMATE

Mohd Talha Anees[*], *PhD*
and Ahmad Farid Bin Abu Bakar, PhD
Department of Geology, Faculty of Science, University of Malaya,
Kuala Lumpur, Malaysia

ABSTRACT

Wind speed forecasting is one of the most relevant and challenging world research problems nowadays. It is related mostly to wind disaster forecasting, wind energy, coastal flood forecasting, other related issues. Several methods were developed for wind speed forecasting in different sectors and climates. The focus of this chapter will on the relationship between wind speed forecasting and wind disasters. There are a number of wind disasters based on the magnitude of wind speed and their origin. Wind speed and direction play an important role in regulating a wide variety of meteorological phenomena. A review of recent studies on

---

[*] Corresponding Author's Email: talhaanees_alg@yahoo.in.

accurate wind speed forecasting from different data sets, models, and algorithms will be the main aim of this chapter. This chapter also discussed the classification of extra-tropical and tropical cyclones because of their dynamic origins. Furthermore, the role of remote sensing in wind speed estimation was also reviewed.

Keywords: wind speed forecasting, extra-tropical cyclone, tropical cyclone, remote sensing

INTRODUCTION

Overview of Wind Motion and Their Affecting Factors

The wind is simply a horizontal movement of air relative to the earth's surface caused by atmospheric pressure differences. Wind speed and direction are important for weather forecasting and monitoring global climate change. There are a number of factors affecting wind motion operating from micro to macro scale in the world. First, pressure difference or pressure gradient, either in the atmosphere or on the earth's surface, in which wind moves from higher pressure to lower pressure. Second, the Coriolis force due to the rotation of the earth which is zero at the equator and maximum at the poles. It acts at right angles to the horizontal direction of the wind (Lal, 2009). Third, frictional forces act at or near (a few thousand meters from the surface) the earth's surface which is responsible for slowing downwind speed and changing wing direction. Fourth, centrifugal action of wind which acts on the circular motion of air caused by Coriolis force. Fifth, local weather conditions which are responsible for changing wind speed due to the formation of wind disasters and monsoons. These are the summary of general factors affecting wind speed and direction and creates different types of situations and natural hazards in the world.

Types of situations or conditions for wind motion created by variations in pressure and temperature which depends on the amount of insolation received at the earth's surface. The distribution of insolation at the earth's surface is unequal on the basis of some astronomical and geographical factors such as the angle of incidence, duration of sunshine, solar constant, the distance between the earth and the sun, and transparency of the atmosphere. Heating of a particular place at the earth's surface reduces pressure and expends the volume of the air in that particular area while cooling of the atmosphere creates vice versa situation. Heating and cooling of the atmosphere depend on (i) partial absorption of solar radiation by the atmosphere, (ii) conduction, (iii) terrestrial radiation, (iv) convection and advection, (v) latent heat of condensation, and (vi) expansion and compression of the air. Apart from latitude and longitude, the temperature at the earth's surface is controlled by the distribution of land and water, movement of ocean currents, prevailing winds, cloud cover, nature of topography such as mountain barriers and variation in reliefs, nature of the surface in terms of color distribution, and convection and turbulence in the atmosphere. Some of the above-mentioned factors operate locally and some are global which produces different situations based on different scenarios for Spatio-temporal temperature and pressure distribution in the world.

Overview of Temperature and Pressure Distribution in the World

Variations in weather conditions are closely associated with temperature and pressure differences. The general distribution of pressure belts are divided into seven alternating low and high-pressure belts on the earth's surface (Lal, 2009): (i) equatorial low-pressure belt: It is located between 5°N and 5°S latitude where heating is intense due to high temperature throughout the year. Trade wind from subtropical high-pressure belts in the northern and southern hemispheres converge here due to continuous low pressure in this belt, (ii) sub-tropical high-pressure belt (in the northern and southern hemisphere): It is located between 25°N and 35°S latitudes and broken into a number of high-pressure centers. In these regions, air sinks and settles from higher altitudes cause fair and dry air throughout the year and increment in surface air temperature. All the hot deserts of the world lie in this belt and hence it is the region of stormy winds (Cherchi et al., 2018), (iii) subpolar low-pressure belts (in the northern and southern hemisphere): It is located between 60°N and 70°S latitudes. The southern belt is an uninterrupted low-pressure belt while it is broken in the northern hemisphere due to the presence of large landmasses, and (iv) polar highs (in northern and southern hemispheres): In these regions, high pressure is throughout the year. In the southern hemisphere, high pressure is centered at the pole whereas, in the northern hemisphere, it is believed to extend from northern Greenland to westward across the islands situated in the northern part of Canada (Lal, 2009). Because of pressure differences in different regions, several types of wind disaster originates which are associated with a particular situation of a particular region.

NATURAL HAZARDS ASSOCIATED WITH WIND SPEED

Atmospheric disturbances due to uneven distribution of low and high-pressure belts lead to creating different types of wind system at the earth's surface. Circulation of wind can be classified into three broad categories (Lal, 2009): Primary circulation which include wind system associated with types of pressure belts such as trade winds (developed due to pressure gradient from subtropical high-pressure belt to the equatorial low-pressure belt), westerlies (movement from subtropical high pressure towards poles), and polar easterlies (movement from polar high to sub-polar low-pressure belt). Secondary circulations include the development of cyclones, anticyclones, monsoons, and air masses. Tertiary circulations include all the local winds and affect local weather conditions. The general explanation of spatial atmospheric disturbances will be helpful to understand the origin of several natural hazards associated with wind speed.

Cyclones

A cyclone is rapid inward circulation of an air mass around a low-pressure center, anticlockwise and clockwise circulations in northern and southern hemisphere respectively, due to atmospheric disturbances (Marchigiani et al., 2013). Cyclone can be divided into two broad categories: (i) extra-tropical cyclone and (ii) tropical cyclone. Extra-tropical cyclones originate from the polar front and dominant in a mid-latitude where contrasting air masses generally meet. Traveling of this cyclone is associated with strong winds, precipitation, and change in temperature (Ulbrich et al., 2009). Generally, the diameter of the extra-tropical cyclone ranges from 300 km to 1500 km. In these cyclones, the pressure is low at the center (vary from 930 mb to 1000 mb) while increasing towards the margin (vary from 10 to 20 mb). Tropical cyclones originate between sub-tropical high-pressure belts between the northern and southern hemispheres. These cyclones named according to their origin as hurricane or typhoons. Hurricanes referred to as those cyclones that arise in the North Atlantic and the eastern North Pacific oceans while those that arise in the Indian and South Pacific Oceans are referred to as typhoons. Wind speed of hurricane exceeds 120 km per hour to 200 km per hour which is the destructive wind speed. The pressure gradient is always steep in hurricanes and it moves from the equator towards the west after their formation. It is the most violent and devastating natural disaster due to atmospheric disturbances and caused widespread damages over land areas (Xi et al., 2019).

Tornadoes

Tornado is a short duration and small in size low-pressure centers having a whirlpool-like structure of winds rotating around a central cavity where a partial vacuum is produced by the centrifugal forces (Lal, 2009). Wind speed, revolving tightly around the core, exceeds from 300 km per hour and it travels over the ground with speed 30 to 45 km per hour. It develops anywhere in the world except extremely cold northern parts of the continents during winters and the Polar Regions.

WIND SPEED FORECASTING METHODS

Wind disasters originate from oceans and then moves towards the land where they affect the coastal areas through powerful winds and torrential rain (Wei, 2015). Strong winds produce large waves with enough energy while bellowing over the sea to influence marine structures and erode beaches (Chang and Chien, 2006). For this purpose, accurate wind speed forecasting models/methods need to be developed for better disaster prediction and hazard management. There are four types of approaches for wind speed forecasting based on temporal range mentioned in the literature (de Freitas et al., 2018): (i) long-term (from a week to year or much ahead), for forecasting a series of multiple events; (ii) medium-term (from two days to a week ahead); (iii) short-term (from one hour to two days), for forecasting due to single event; and (iv) very short-term (from a few seconds to one hour). However, these approaches are very common in wind energy prediction for wind turbines. Table 1 showing scales of wind motion in the atmosphere which can correlate with these approaches to understand their applications.

These approaches have been implemented using three general types of wind speed forecasting models such as physical model, statistical model, and hybrid model. Physical models are realistic mathematical models that integrate a large amount of historical physical data such as barriers, temperature, pressure, and wind characteristics obtained from numerical weather prediction (NWP) (de Freitas et al., 2018). NWP solves a set of partial differential equations for large scale forecasting that govern atmospheric motion and evolution (Kalnay 2003) and provide accurate Spatio-temporal weather prediction.

Statistical models, unlike physical models, delineate input and output data relationship for detecting patterns in databases, through analysis of a massive training dataset. Statistical models have been used for short-term predictability of wind speeds using NWP, stochastic and hybrid models (Ferreira et al., 2019). Statistical models further sub-divided into three categories such as time-series models, spatial correlation, and intelligence artificial (IA) methods (de Freitas et al., 2018).

Hybrid models for wind speed forecasting consist of linear and non-linear models to improve the performance of wind speed prediction (Vidya and Janani, 2019). Artificial Intelligence methods (AI) and hybrid models are state of the art approaches for accurate wind speed prediction and provide best results in most of the cases (de Freitas et al., 2018). However, hybrid structures provide relevant results and good performance compared to AI methods.

Table 1. Scales of wind motion in the atmosphere

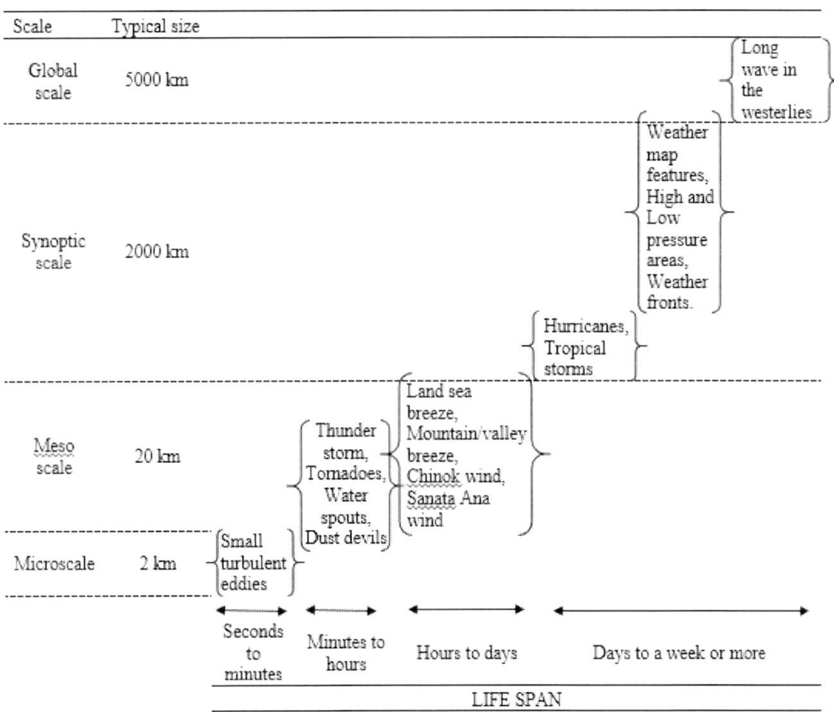

In the following section, several case studies on the above mentioned different approaches and models on wind speed forecasting for wind associated natural hazards of different climates will be presented. Secondary hazards due to wind disasters and the role of anthropogenic threats to the coastal zones will also be analyzed. Furthermore, the role of remote sensing and GIS in wind speed forecasting will be discussed.

CASE STUDIES

Classification of Extra-Tropical Cyclones

Extra-tropical cyclones are important for mid-latitude weather. Due to the dynamic behavior of atmospheric disturbances, extra-tropical cyclones can be classified using several approaches. Recently, Catto (2016) review the extra-tropical cyclone classification approaches and its use in the state of the art climate models for their realistic representation. The summary of different classification are as follows (Catto, 2016):

Classification Based on Simple Conceptual Models and Life Cycle Characteristics

(i) The development of conceptual models: Two models such as the Norwegian model and the Shapiro-Keyser model were discussed which were developed to describe cyclone genesis and development. Both models describe the different type of separation procedure of warm and cold fronts.

(ii) Idealized modeling: It is based on the development of theories for the evolution of extratropical cyclones, their associated fronts, and their impact. Under this classification, two life cycles were developed which were differed in the amount of cyclonic shear in the initialization. These two life cycles mentioned as useful tools that can represent the large-scale and smaller-scale features of extratropical cyclogenesis.

(iii) Explosive cyclones: Classification of cyclones based on their rapid deepening or "explosive cyclogenesis" and can be defined by their central pressure which must be at least 1 Bergeron. These types of cyclones can have large impacts due to their precipitation, storm surge, or strong winds. Other factors depending on their location are upper-level forcing, air-sea interactions, and diabatic heating.

Classification Based on Satellite Images

The satellite classification of the extra-tropical cyclone is based on their distinctive cloud signatures from which upper-level flow features can be obtained. This feature is a very powerful tool for understanding extratropical cyclogenesis when it combined with analysis data. Satellite classification was further sub-divided based on different upper-level flow features which were mentioned in detail in that review.

Classification Based on Airstream Analysis

This classification was based on the ideas of cyclone-relative airflows and further sub-divided into the warm conveyor belt, the cold conveyor belt, and the dry intrusion.

Classification Based on Synoptic-Dynamic

(i) Upper Level versus Lower Level Forcing: This classification was based on two mechanisms such as the baroclinic environment of the polar front and positive vorticity advection in the upper troposphere. This classification is further sub-divided into three types A, B, and C based on a ratio of thermal advection at low levels and vorticity advection at upper levels.

(ii) Secondary cyclones: These cyclones are classified based on frontal and non-frontal cyclones and their intensities.

(iii) Upper-level flow features: It was based on upper-level flow configurations at the time of cyclogenesis

(iv) Other cyclone features: classification based on other features such as cyclone intensity and shapes of pressure patterns.

Classification Based on Transitioning Extratropical Cyclones

This classification was based on the transition of the extra-tropical cyclone from tropical cyclone when it interacts with the mid-latitude westerlies and increases its translation speed. This type of cyclones associated with intense wind and heavy rainfall and could have a large socio-economic impact.

Classification Based on Sub-Tropical Cyclone and East Coast Lows (ECL)

Generally, sub-tropical environmental condition is not suitable for the development of cyclone. Extra-tropical cyclone also not able to develop in sub-tropic due to weak baroclinicity. However, there is sufficient vertical wind shear, and latent heating due to convection, subtropical cyclones can form even without strong surface baroclinicity. ECL is also considered as an extratropical cyclone that occurred in cool-season of east Australia. ECL has strong winds, high seas, and heavy precipitation.

Classification Based on Polar Lows

Polar lows are mesoscale and intense cyclones that typically form when very cold air passes over relatively warmer seas.

Classification Based on Impacts

Further classification of extra-tropical cyclones were based on the impact of strong winds, precipitation and clustering.

Case Studies of Extra-Tropical Cyclones

Brâncuş et al. (2019) analyzed the evolution of strong winds using a mesoscale model simulation to investigate the physical processes that were responsible for the strong winds and their acceleration at the Black Sea and east coast of Romania. The extratropical cyclone was identified as the explosive type as an atmospheric bomb by deepening 1.5 Bergerons and it followed the Shapiro–Keyser model and developed strong winds equatorward of the cyclone center. With heavy rainfall, strong wind gusts were recorded up to 36 m/s at Medgidia (over land) and up to 38 m/s at the Gloria Oil Platform (over sea) for a period of about 4–6 h. Two models were used in this study such as the Advanced Research Weather Research and Forecasting (WRF) Model and The European Centre for Medium-Range Weather Forecasts. The model was run for 36 hours with output at every 15 minutes to capture the detail of the structure and evolution of the wind field during this time. Petterssen frontogenesis was used to understand the responsible mechanisms for the generation of strong winds. To investigate the acceleration of strong winds, a horizontal momentum equation was used. The results showed that the structure of wind was tear-shaped and the strongest winds were located in high relative humidity (80 to 90%) region. The origin of strong winds was two-fold associated with different two different airstreams. Overall, it was concluded that the strong wind had more than one source and successfully forecast strong winds due to extratropical cyclone.

Befort et al. (2019) investigate the extent of two modern seasonal forecast systems such as the ECMWF and theMet Office Hadley Centre and their capabilities of forecasting the frequency of extratropical cyclones and windstorms over the Northern Hemisphere. The skills of windstorm forecasting were assessed using North Atlantic Oscillation (NAO) as a predictor. Also, the relationship between NAO and windstorm frequency was explored. First, the cyclone was tracked using an algorithm mentioned in Simmonds and Murray (1999) and Simmonds et al. (1999). The tracking scheme developed by Leckebusch et al. (2008) for windstorm events were followed. The results indicated that both extratropical cyclones and windstorms are well represented in each of the models, across the Northern Hemisphere with some regional errors. Forecasting skills using NAO showed lower skill along the nodal line of the NAO including most central-western European countries when compared to the direct approach (where actual windstorm events are directly identified from wind speeds).

Hewson and Neu (2015) developed a conceptual model of windstorm generating cyclone by analyzing 29 historic windstorm data. The mesoscale structure of extra-tropical cyclonic windstorms and their representation in re-analysis data were investigated. Footprints damages were examined by three wind phenomena such as the warm jet (WJ), the sting jet (SJ), and the cold jet (CJ) which were delineated by the conceptual model. Overall results showed that WJ developed early in the life cycle while SJ is the smallest and short-lived but most damaging. To capture the most damaging winds, a horizontal resolution of 20 km or better was suggested based on results.

Studholme et al. (2015) presented a method to combine numerical and computational models with the characterization of storm structure. K-means clustering was applied to isolate different life-cycle stages of cyclones. The proposed methodology was tested from the European Centre of Medium-Range Weather Forecasting (ECMWF) operational analyses. Results showed that 54% of Northern Hemisphere's tropical storms undergo an extratropical transition (EL) in that given year. However, the error presented from this analysis was not explained without any reason. Overall, the methodology is effective for determining ET transition.

Classification of Tropical Cyclones

Tropical cyclones differ from one region to another. It is called from a different name based on the origin, wind forces, wind scales and speeds, and damages. Chenoweth (2007) analyzed newspaper reports for a period of 1795 to 1879 and classified tropical cyclones on the basis of wind force terms and wind damage on land when the location, speed, and direction of movement of the tropical cyclone. Cullen (2002) classified tropical cyclones on the basis of tree damages and wind speed and scales. Several classifications also published by the World Meteorological Organization (WMO) in different reports. Table 2 showing tropical cyclone classification on the basis of 10-minute sustained winds used by WMO.

Table 2. Classification of tropical cyclone (WMO, 2008; CPHC, 2019; Nam-Young and Elsner, 2018; IMD, 2018)

10 min sustained winds (WMO)	North Eastern Pacific & North Atlantic (NGC/CPHC)	North Western pacific (JMA)	North Indian Ocean (IMD)	South Indian Ocean (MF)	Australia & South Pacific (BOM/FMS)
< 28 knots (32 mph; 52 km/h)	Tropical Depression	Tropical Depression	Depression	Zone od Disturbed Weather	Tropical Disturbance Tropical Depression Tropical Low
28–29 knots (32–33 mph; 52–54 km/h)			Deep Depression	Tropical Disturbance	
30–33 knots (35–38 mph; 56–61 km/h)	Tropical Strom			Tropical Depression	
34–47 knots (39–54 mph; 63–87 km/h)		Tropical Storm	Cyclonic Storm	Severe Tropical Storm	Category 1 Tropical Cyclone
48–55 knots (55–63 mph; 89–102 km/h)		Severe Tropical Storm	Severe Cyclonic Storm		Category 2 Tropical Cyclone
56–63 knots (64–72 mph; 104–117 km/h)	Category 1 Hurricane			Tropical Cyclone	
64–72 knots (74–83 mph; 119–133 km/h)		Typhoon	Very Severe Cyclonic Storm		Category 3 Severe Tropical Cyclone
73–83 knots (84–96 mph; 135–154 km/h)	Category 2 Hurricane				
84–85 knots (97–98 mph; 156–157 km/h)	Category 3 Major Hurricane	Very Strong Typhoon		Intense Tropical Cyclone	
86–98 knots (99–113 mph; 159–157 km/h)			Extremely Severe Cyclonic Storm		Category 4 Severe Tropical Cyclone
99–107 knots (114–123 mph; 183–198 km/h)	Category 4 Major Hurricane				
108–113 knots (124–130 mph; 200–209 km/h)		Violent Typhoon		Very Intense Tropical Cyclone	Category 5 Severe Tropical Cyclone
114–119 knots (131–137 mph; 211–220 km/h)			Super Cyclonic Storm		
> 120 knots (138 mph; 222 km/h)	Category 5 Major Hurricane				

Case Studies of Tropical Cyclones

Lee and Rosowsky (2007) developed a database of synthetic hurricane wind speeds for a region of the south-eastern United States. The database includes a time index (year), duration of strong winds (greater than 10 m/s), and peak surface wind speeds (both gust and sustained). Two models such as the short-term and long-term wind speed database were used to calibrate damage models and in hurricane risk analyses respectively. Also, some sub-models such as the gradient wind field model, tracking model, central pressure model, decay model, and boundary layer model were used to improve the model performance. Short-term wind speed database was also used to predict hurricane wind speeds and damage patterns in advance of approaching storms. Whereas, the long-term wind speed database was used for design wind speed maps, or for emergency management and preparedness activities.

DeMaria et al. (2009) proposed the new wind probability model including the uncertainties in the track, intensity, and wind structure forecasts, and estimates the probabilities of 34-, 50-, and 64-kt winds out to 120 h for all tropical cyclones in the Northern Hemisphere. Monte Carlo method was used where 1000 realizations are generated by randomly sampling from the operational track and intensity forecast error distributions from the past 5 yr. The results showed that the average probability error was, 0.6% and the maximum error anywhere in the domain was, 4% for all three wind speed thresholds. The proposed model was relatively unbiased and the forecasts were skillful based on the Brier skill score and the relative operating characteristic skill score. Furthermore, the model probabilities are also well-calibrated and have high confidence based on reliability diagrams.

Holland et al. (2010) presented a revised approach for the parametric radial profile of hurricane winds by allowing a variation in the wind equation exponent to fit inner and outer wind observations. The data was obtained from hurricane archives or in hurricane warning information and from some existing parametric models of the hurricane surface wind field. Input data consist of an outer wind and surface pressure, sea surface temperature, radius of maximum winds, and central pressure. Secondary wind maxima were included by adding a perturbation with specified amplitude and radial range to the standard profile. Results showed that the proposed model was dynamically stable with a bias of less than 10% and demonstrated to be an improvement on two other wind profile models.

Ishihara and Yamaguchi (2015) forecast extreme wind speed of tropical cyclones using the Monte Carlo simulation (MCS) and measure-correlate-predict (MCP) method. The models were validated by generating synthetic tropical cyclones. Uncertainties of extreme wind speed estimated by MCS from a limited number of annual maximum wind speed and the limited length of the pressure measurement data used for the estimation of the tropical cyclone parameters. Also, the extreme wind speed was estimated by including these uncertainties. The results showed that larger uncertainties were reported for the extreme wind speed induced by a tropical cyclone, which implies the possibility of underestimation. The proposed MCS model worked well for extreme wind speed prediction of a tropical cyclone.

Wei (2015) forecast the hourly wind speeds (up to 6 hours) over two offshore islands near Taiwan such as Lanyu, and Pengjia Islets, during tropical cyclones using machine learning techniques. The four kernel-based support vector machines for regression (SVR) models, comprising radial basis function, linear, polynomial, and Pearson VII universal kernels were used to develop a highly reliable surface wind speed prediction technique. To maintain the accuracy of the SVR model, traditional regressions and the parametric wind representations, comprising the modified Rankine profile, Holland wind profile, and DeMaria wind profile were used to compare wind speed forecasts. A total of 84 typhoon events that occurred along the coast of the western North Pacific region over the past 13 years were included in this study. These typhoons were developed due to the increment of hurricane wind speed which exceeds the intensities of tropical depressions and tropical storms. Wind intensities were varied from less than 18 m/s (tropical depression) to 70 m/s (Category 5 according to the Saffir-Simpson wind scale). The results indicated that the Pearson VII SVR is the most precise of the kernel-based SVR models, regressions, and parametric wind representations. However, Holland wind profile seems unsuitable for areas affected by topographic effects, such as the Central Mountain Range of Taiwan but applicable to open ocean

Haghroosta (2019) addressed the uncertainties in predicting typhoon wind speed in the South China Sea. The performance of Weather Research and Forecasting (WRF), and an Adaptive Neuro-Fuzzy Inference System (ANFIS) model are calculated using statistical parameters of the root mean square error (RMSE) and Correlation Coefficient (CC), and the advantages and disadvantages of both models are represented. The results showed that ANFIS can predict typhoon intensity in the study area, but it needs too much historical data to achieve better answers. Whereas WRF needs high capacity in terms of computational technology and has too many complicated calculations and interpolations that occur during the pre-process, running, and post-process of the WRF model that can create more inaccuracies in the outcomes. Other disadvantages of WRF is approximate predictions and dependency on the purpose of the model, real-time versus research, type of domain and its resolution, the types of variables associated to the study objectives, how to run the model, and model run-time. ANFIS had an advantage over WRF on accuracy in typhoon intensity prediction. The neural network model (a non-linear statistical model) has the advantage of less formal statistical training and the capability of identifying complex nonlinear correlations between dependent and independent variables by various learning algorithms. Whereas, the disadvantage is the nature of the "black box" that unable to interpret the relationship between input and output data.

Role of Remote Sensing in Wind Speed Forecasting

QuikSCAT (QS): The SeaWinds scatterometer on QS satellite, launched in June 1999, is a scanning microwave radar that measures electromagnetic backscatter from the wind roughened ocean surface at multiple antenna look angles which is used to infer the surface wind stress magnitude and direction (Jelenak et al., 2009). QS Samples 90% of the global oceans daily and provided the longest and highest quality remotely-sensed global ocean surface wind vector time-series to date.

Advanced Scatterometer (ASCAT): ASCAT MetOp-A and MetOp-B were launched in October 2006 and in September 2012 respectively to measure wind speed and direction over the oceans for studying polar ice, soil moisture and vegetation. Zabolotskikh et al. (2014) compared the results between sea surface wind speed estimation over North Atlantic from MetOp-A and GCOM-W1 Advanced Microwave Sounding Radiometer 2 (AMSR2). Results showed that estimation from both was highly correlated with each other.

X-Band Radar: X-band is a segment of the microwave radio region of the electromagnetic spectrum, within the frequency range 8–12 GHz mounted on most large research vessels for ship traffic control and navigation and also waves and current measurement. Wind speed of more than approximately 3 m/s, the backscatter from the sea surface becomes visible in radar images. X-band radar systems scan the ocean surface in real-time at high temporal (1–2 s) and spatial (5–10 m) resolution. Huang and Wang (2016) mentioned the wind speed retrieval method from X band marine radar images using algorithms. Chen et al. (2018) proposed a support vector regression (SVR) based method for estimating wind speed from X-band marine radar images.

The Stepped Frequency Microwave Radiometer (SFMR): SFMR is an aircraft-based, nadir-looking microwave radiometer developed by ProSensing Inc. of Amherst, MA, USA (Sapp et al., 2019). Primarily, it was used to determine hurricane intensities by the National Hurricane Centre, Miami, Florida. SFMR measures temperatures brightness (Tbs) of the ocean surface and intervening atmosphere at six C-band frequencies. From these Tbs, surface wind speed and average columnar rain rate can be estimated (Sapp et al., 2019).

Kent et al. (2019) used multiple remote sensing data for wind speed estimation. Global digital elevation models (GDEM): Advanced Spaceborne Thermal Emission and Reflection Radiometer (ASTER), Shuttle Radar Topography Mission (SRTM), and TanDEM-X were used to derive urban morphology and aerodynamic roughness parameters. Using these roughness parameters, wind speeds were estimated up to ~10 times canopy height from a reference observed wind speed at ~2.5 times canopy height. The results showed that 40% underestimation of wind speed was from the ASTER and SRTM and 30% from the TanDEM-X.

CONCLUSION

This study review recent studies on wind speed forecasting methods for wind disaster mainly extra-tropical and tropical cyclones. Because of the dynamic origins of extra-tropical and tropical cyclones, their classifications were also discussed from recent studies. At the end of this chapter, it is concluded that several models or methods have been used for wind speed forecasting in different cyclones. Each mode/methods have their advantages and disadvantages. However, artificial neural network and machine learning methods using big data have grater accuracies over other methods. The role of remote sensing in Spatio-temporal wind speed and temperature data collection is very important and plays a crucial role in future studies. Based on this review, it is recommended that an integrated approach of different models/methods will be useful for a cyclone of dynamic origins and high-resolution remote sensing data will be useful for accurate Spatio-temporal wind speed data measurement.

REFERENCES

Befort, D. J., Wild, S., Knight, J. R., Lockwood, J. F., Thornton, H. E., Hermanson, L., Bett, P. E., Weisheimer, A. & Leckebusch, G. C. (2019). Seasonal forecast skill for extratropical cyclones and windstorms. *Quarterly Journal of the Royal Meteorological Society*, *145*(718), 92-104.

Brâncuş, M., Schultz, D. M., Antonescu, B., Dearden, C. & Ştefan, S. (2019). Origin of strong winds in an explosive mediterranean extratropical cyclone. *Monthly Weather Review*, *147*(10), 3649-3671.

Chang, H. K. & Chien, W. A. (2006). Neural network with multi-trend simulating transfer function for forecasting typhoon wave, *Adv. Eng. Software*, *37*, 184–194.

Chen, X., Huang, W. & Yao, G. (2018). Wind speed estimation from x-band marine radar images using support vector regression method. *IEEE Geoscience and Remote Sensing Letters*, *15*(9), 1312-1316.

Chenoweth, M. (2007). Objective classification of historical tropical cyclone intensity. *Journal of Geophysical Research: Atmospheres*, *112*(D5).

Cherchi, A., Ambrizzi, T., Behera, S., Freitas, A. C. V., Morioka, Y. & Zhou, T. (2018). The response of subtropical highs to climate change. *Current Climate Change Reports*, *4*(4), 371-382.

CPHC. (2018). Regional Association V - Tropical Cyclone Operational Plan for the South Pacific and South-East Indian Ocean, Tropical Cyclone Programme Report No. TCP-24, Geneva 2, Switzerland.

Cullen, S. (2002). Trees and wind: wind scales and speeds. *Journal of Arboriculture*, *28*(5), 237-242.

de Freitas, N. C., Silva, M. P. D. S. & Sakamoto, M. S. (2018). *Wind Speed Forecasting: A Review.*, *8*(1), 4-9.

DeMaria, M., Knaff, J. A., Knabb, R., Lauer, C., Sampson, C. R. & DeMaria, R. T. (2009). A new method for estimating tropical cyclone wind speed probabilities. *Weather and Forecasting*, *24*(6), 1573-1591.

Ferreira, M., Santos, A. & Lucio, P. (2019). Short-term forecast of wind speed through mathematical models. *Energy Reports*, *5*, 1172-1184.

Harper1, B. A., Kepert, J. D. & Ginger, J. D. (2008). World Meteorological Organization (WMO), guidelines for converting between various wind averaging periods in tropical cyclone conditions.

Hewson, T. D. & Neu, U. (2015). Cyclones, windstorms and the IMILAST project. *Tellus A: Dynamic Meteorology and Oceanography*, *67*(1), 27128.

Holland, G. J., Belanger, J. I. & Fritz, A. (2010). A revised model for radial profiles of hurricane winds. *Monthly Weather Review*, *138*(12), 4393-4401.

Huang, W. & Wang, Y. (2016). A spectra-analysis-based algorithm for wind speed estimation from X-band nautical radar images. *IEEE Geoscience and Remote Sensing Letters*, *13*(5), 701-705.

IMD (Indian Meteorological Department). (2018). Tropical cyclone operational plan for the Bay of Bengal and the Arabian Sea, Tropical cyclone programme Report no. TCP-21 World Meterological Organization Technical Document, Geneva 2, Switzerland.

Ishihara, T. & Yamaguchi, A. (2015). Prediction of the extreme wind speed in the mixed climate region by using Monte Carlo simulation and measure-correlate-predict method. *Wind Energy*, *18*(1), 171-186.

Jelenak, Z., Ahmad, K., Sienkiewicz, J. & Chang, P. S. (2009, July). A statistical study of wind field distribution within extra-tropical cyclones in North Pacific ocean from 7-years of QuikSCAT wind data. In *2009 IEEE International Geoscience and Remote Sensing Symposium*, (Vol. *1*, pp. I-104). IEEE.

Kalnay, E. (2003). Atmospheric Modeling, Data Assimilation, and Predictability (Cambridge University Press, 2003).

Kent, C. W., Grimmond, S., Gatey, D. & Hirano, K. (2019). Urban morphology parameters from global digital elevation models: Implications for aerodynamic roughness and for wind-speed estimation. *Remote sensing of environment*, *221*, 316-339.

Leckebusch, G. C., Renggli, D. & Ulbrich, U. (2008). Development and application of an objective storm severity measure for the northeast Atlantic region. *Meteorologische Zeitschrift*, *17*, 575–587. https://doi.org/10.1127/0941-2948/2008/0323.

Lee, K. H. & Rosowsky, D. V. (2007). Synthetic hurricane wind speed records: development of a database for hazard analyses and risk studies. *Natural Hazards Review*, *8*(2), 23-34.

Marchigiani, R., Gordy, S., Cipolla, J., Adams, R. C., Evans, D. C., Stehly, C., Galwankar, S., Russell, S., Marco, A. P., Kman, N. & Bhoi, S. (2013). Wind disasters: A comprehensive review of current management strategies. *International journal of critical illness and injury science*, *3*(2), 130.

Nam-Young, K. & Elsner, J. B. (2018). The changing validity of tropical cyclone warnings under global warming. *NPJ Climate and Atmospheric Science*, *1*(1).

Sapp, J. W., Alsweiss, S. O., Jelenak, Z., Chang, P. S. & Carswell, J. (2019). Stepped frequency microwave radiometer wind-speed retrieval improvements. *Remote Sensing*, *11*(3), 214.

Simmonds, I. & Murray, R. J. (1999) Southern extratropical cyclone behaviour in ECMWF analyses during the FROST special observing periods. *Weather and Forecasting*, *14*, 878–891. https://doi.org/ 10.1175/1520-0434(1999)014<0878:SECBIE>2.0.CO;2.

Simmonds, I., Murray, R. J. & Leighton, R. M. (1999). A refinement of cyclone tracking methods with data from FROST. *Australian Meteorological Magazine*, (Special Issue), 35–49.

Studholme, J., Hodges, K. I. & Brierley, C. M. (2015). Objective determination of the extratropical transition of tropical cyclones in the Northern Hemisphere. *Tellus A: Dynamic Meteorology and Oceanography*, *67*(1), 24474.

Ulbrich, U., Leckebusch, G. C. & Pinto, J. G. (2009). Extra-tropical cyclones in the present and future climate: a review. *Theoretical and Applied Climatology*, *96*(1-2), 117-131.

Vidya, S. & Janani, E. S. V. *A Review On The Hybrid Approaches For Wind Speed Forecasting*, *8*(9), 1584-1590.

Wei, C. C. (2015). Forecasting surface wind speeds over offshore islands near Taiwan during tropical cyclones: Comparisons of data-driven algorithms and parametric wind representations. *Journal of Geophysical Research: Atmospheres*, *120*(5), 1826-1847.

Xi, W., Peet, R. K., Lee, M. T. & Urban, D. L. (2019). Hurricane disturbances, tree diversity, and succession in North Carolina Piedmont forests, USA. *Journal of forestry research*, *30*(1), 219-231.

Zabolotskikh, E., Mitnik, L. & Chapron, B. (2014). GCOM-W1 AMSR2 and MetOp-A ASCAT wind speeds for the extratropical cyclones over the North Atlantic. *Remote Sensing of Environment*, *147*, 89-98.

In: Wind Speed: An Overview   ISBN: 978-1-53618-412-9
Editor: Nicolas Koči   © 2020 Nova Science Publishers, Inc.

Chapter 3

STATISTICAL COHERENCE AMONG SOLAR IRRADIATION, AIR TEMPERATURE AND WIND SPEED

*Nurdan Karapinar, Numan Sabit Cetin
and Melih Soner Celiktas**
Solar Energy Institute, Ege University, Izmir, Turkey

ABSTRACT

The working principle of solar panels is the conversion of solar irradiation falling on the entire surface of the panel, to electricity. Conversion efficiency is acquired by capturing the electrical current generated when the sunlight interacts with silicon, multi-junction, GaAs, thin-film, perovskite or organic cells inside the solar panel. Although, this efficiency varies based on the specific features and size of the panels, the typical ratios are between 12 and 47.2% (NREL). All of these calculations are based on the estimation of the amount of solar energy from the sun under current seasonal conditions in order to achieve an accurate result and to obtain maximum efficiency by the solar panels.

Wind energy has taken an important place within the clean energy sources. The amount of energy that can be produced from a wind turbine

---

* Corresponding Author's Email: soner.celiktas@ege.edu.tr.

is directly related to the value of the wind speed in that specific location. Therefore, it is very important to determine the wind speed accurately for each region. Besides that, with the transition to smart grids, the inclusion of different production resources into the grid has revealed the necessity of managing the grid efficiently.

In this study, data from the Modern-Era Retrospective analysis for Research and Applications, version-2 (MERRA-2) software [1] delivering time series of air temperature, local solar irradiation and wind speed were used for further calculations. For in situ measurement of the ground data, the measuring device was placed at an altitude of 50 m at the outdoor premises of the Institute of Solar Energy at Ege University, Izmir, Turkey. A statistical coherence among solar irradiation, air temperature and wind speed were observed and expressed as a third degree polynomial equation. With the equation developed, the differences between satellite data and ground data were corrected. Then, the differences between the corrected temperature and solar data were plotted. The value of the determination coefficient (R^2) in the resulting graphs has given the accuracy of the improved data. With this improvement, similar values have been achieved (approximately with 25% error for temperature and about 15% error for solar radiation and 0.0012% error for wind speed) using satellite data without ground measurement data. For this purpose, a different formulation was developed to obtain the results of a location-based measurement center from the satellite measurement results. This coherence obtained from solar irradation can be utilized to predict the changes over the years and perform calculations of e yield analysis.

Keywords: solar energy, solar irradation, wind speed, air temperature

INTRODUCTION

Renewable energy use has attracted great interest from investors and researchers in recent years due to the availability and sustainability of wind and solar energy. In the last few decades, many wind turbines and photovoltaic fields [2, 3] have been built. The harmony between solar energy and temperature are extremely important for the performance ratio value of these systems and the systems to be installed later. It is possible to utilize solar energy systems for heating, cooling and electricity generation [4].

Energy is an important part of our daily lives. In parallel with the needs of today's increasing power demand, clean energy production is of great importance. Therefore, the use of renewable energy sources is rapidly increasing. As a result, the share of renewable energy in electricity generation is expected to reach 20% by 2020 and the main part of this renewable energy will be allocated to wind energy, which should be 12% [5]. Photovoltaic energy has become a popular source of renewable energy for sustainable urban development [6]. At local scale, solar radiation can be accurately measured by ground-based meteorological observation stations. However, the presence of observed solar radiation measurements has been spatially and temporarily insufficient for many applications [7]. Uninterrupted and strong wind speeds are required for uninterrupted and high quality electricity production of wind turbines. However, the variability of wind speed brings some important problems. To overcome these problems, short-term wind speed estimation is required. For example, short-term wind speed estimation is required for safe operation (control) of wind fields and turbines. In addition, a reliable short-term wind speed estimate is required for the safe integration of wind energy into the electricity grid and the economic distribution of the network.

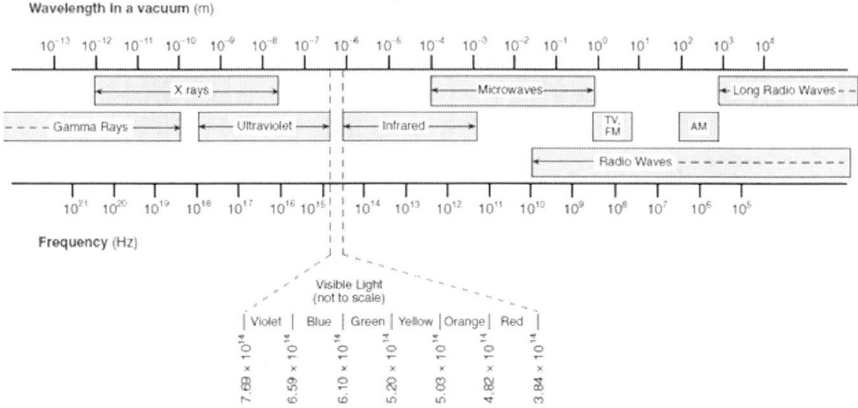

Figure 1. Diagram of the electromagnetic spectrum.

It is a product of all electromagnetic waves arranged according to the electromagnetic spectrum, frequency and wavelength (Figure 1). The sun is a celestial body that emits electromagnetic energy at different wavelengths. Electromagnetic energy moves at the speed of light in sinusoidal waves from space. Light is a specific type of electromagnetic radiation that can be seen and perceived by the human eye, but this energy is available in a wide variety of wavelengths [8]. The visible light and heat energy is a part of the sun that makes life possible, and is named as "sunbeam" or "daylight." This radiation is a fundamental input for many aspects of science. In addition, it is an important parameter in solar energy applications in electricity generation and in day lighting [9].

Various methods are available in the literature for wind speed estimation. Physical atmospheric methods have been accepted for long-term wind speed estimation. For instance, time series based linear regression methods [10, 11, 12], nonlinear Artificial Neural Network (ANN) models [13, 14], Markov chains [15] are the most studied methods. Recently, different hybrid applications [16, 17] have been used in wind speed estimation. On the other hand, some methods that based onneural networks with radial basis function [18, 19], adaptive neuro-fuzzy meaning system, ANN-genetic algorithm, and ANN particle swarm optimization have been used for shor-term forecasting wind speed [20]. The energy coming the Earth's surface in the form of direct or scattered radiation determines the temperature of the inner atmosphere layers of the earth. And, this energy determines the evaporation capacity and climatic features [21]. A mathematical model is used to estimate the received global solar radiation [22]. Some researchers have used kind of methods or models to appraise the solar energy based on include atmospheric transmittance model [23], clearness index model [24] and diffuse fraction model [25]. Temperature is a very random process parameter which is variable both in time and in atmosphere layers. This variability leads to a difficult conversion of energy as solar cells are subjected to an uneven and transient source, varying mechanical loads and non-linear dynamics. In the literature, various distribution functions have been proposed for solar energy prediction modeling and temperature modelling. The short-term wind speed estimate includes estimates over a period of a few seconds to several hours, which can be used to improve the quality of the wind power and ensure the safe integration of the generated energy into the grid [26]. However, it is known that wind speed data are statistically distributed according to certain rules. There are many studies in the literature regarding the estimation of parameters of these distributions and forecasts of short-term wind speeds. According to Masseran [27], the information several from several goodness of fit criteria, the R^2 coefficient, Kolmogorov-Smirnov statistics, Akaike's information criterion and deviation in skewness/kurtosis were integrated for the conclusive selection of the best fit distribution model of wind speed data.

In this study, there are two important differences among the methods in the literature. The purpose of this study is:

1. To be able to create empirical relationships among solar radiation, wind speed, and temperature which can be obtained without the use of field measurements.
2. To acquire knowledge about the instant solar radiation, temperature and wind speed obtained from a particular location.

METHODS

Meteorological data were used to get the range in daily total solar irradiance, air temperature and wind speed in Bornova, Izmir, Turkey on a daily basis between 06 Jan 2013 and 12 Jan 2013. The station of the measurements was located at latitude 38° 27 N, longitude 27° 14 ~ E at an altitude of 50 m. Maximum and minimum temperatures were recorded at screen height. Daily solar irradiance and wind speed incident was collected by means of a pyranometer. In addition, Satallite data of solar irradiance, daily temperature and wind speed were obtained from MERRA-2 on a daily basis between 06 Jan 2013 and 12 Jan 2013.

Solar energy radiates from atmospheric layers radiates, and this form of energy can be kept and used in various ways. Its usage produces less carbon dioxide emissions into the atmosphere thereby not contributing to the excessive pollution created by other sources such as fossil fuels. There

is therefore the growing need to study this type of energy in order to come
out with relevant solutions and practices to harness this type of energy.

The range in daily temperature (T) is calculated as follows:

$$\Delta T(J) = Tmax(J) - (Tmin(J) - Tmin(J+1))/2 \tag{1}$$

where T_{max} is the daily maximum temperature (°C), T_{min} is the daily
minimum temperature (°C), and J is the day number. The average of the
incoming minimum temperature used on both sides of the daily maximum
value was used to reduce the effect of large-scale hot or cold air masses in
the environment. The amount of warm air masses in the atmosphere during
the day T_{max} (J) may affect only the amount of incoming radiation. When
T_{min} (J-1) is not included in the ΔT value, a larger amount of radiation will
be obtained than actually occurred. Although large scaling is valid in
temperate regions, it is not suitable in tropical regions [28]. In high altitude
tropical regions, these effects are minimal, but at high latitudes, the effects
are predicted to be significant. It can be obtained in tropical regions or at
high latitudes, taking average values for periods of 5, 10 or 15 days,
depending on the degree of weather variability. On average, a significant
measure of the direct heating effect of the sun's rays is reached [29]. The
climate is scaled in terms of measurable weather elements. The most
influential of these weather elements are temperature, relative humidity
and, wind. Depending on these, cloudiness influences the amount of solar
radiation that reaches the surface and the transmission of terrestrial
radiation through the atmosphere [30]. Climate is essentially the synthesis
of the weather in the studied location. The climate of a region is associated
with latitude, altitude, and depending on these water bodies, mountains,
and the prevailing wind direction [31]. Majority of these studies aim to
facilitate the weather forecast. Besides, there has been the development of
temperature and solar irradiation calculation methods from past to present.
In this study, two different equations have been developed for the harmony
between solar irradiation data and wind speed-temperature change. In the
developed equations, data measured at 10-minute intervals, daily and
weekly, were used. The data used were compared from the same location

and meteorological station data-satellite data of the same year, and the obtained data were smooth. The equation developed on the smoothed data is used.

$$R^2 = 1 - (First\ Sum\ of\ Errors/Second\ Sum\ of\ Errors) \tag{2}$$

The best measure of how well the experimental data fit a linear curve is the calculated determination coefficient (R^2) in the regression analysis process. R^2 being equal to one, is a proof that the experimental data provides a flawless linear curve. The more data points, the higher the reliability of R^2. If the linear curve passes through all points on the graph, R^2 is equal to one. In this case, the explainable variation is equal to the inexplicable variation. If the points on the graph deviate from the linear curve, the variation not described will be larger and $R^2 < 1$. The correlation coefficient calculated in the regression analysis and whether there is a relationship between two variables is tested [32]. In particular, this linear regression is used to determine how well a line fits a dataset. It is also the fraction of the total variation in y, or, it shows how well a line follows variations in a data set.

RESULTS AND DISCUSSION

In this study, wind speed, temperature and solar irradiation values collected in ten minutes periods from a meteorology station located in the campus area of Ege University were used. In the modeling of these data, both instantaneous average values and linear prediction equations are designed. The linear prediction model was first proposed in 1795 by Gauss [33]. Later this model has been used in many different areas. For instance, to elicit seismic traces in order to determine the presence of oil modeling in geophysics [34] and for the estimation of solar energy [35]. However, there is no one method available for predicting solar irradiation, air temperature, and wind speed. Many similar methods have been developed for meteorological data prediction in the neural network area [36, 37, 38]. The data in this study were examined with the mathematical analysis approach. In this analysis, the data taken at 10-minute intervals daily was obtained from a total of 1008 data in 144 daily data. The main ıbjective was to obtain more reliable ground data with the improvement of R^2 values. The following equations were developed for temperature, wind speed and solar irradiance.

For temperature, equation (3) was developed regardless of weather conditions:

$$T_{smooth} = 0.0008T^2_{satellite} - 0.0909T_{ground} + 279.4 \tag{3}$$

This equation in T_{smooth} smoothed air temperature data, $T_{satellite}$ daily air temperature in MERRA-2 satellite data, T_{ground} daily air temperature in ground data. In this case the coefficients are shown not to be flexible. Because changes in weather conditions (clouds, time, moisture changes, etc.) should be examined in their exchange period for the improvement of the situation taken into consideration.

It is clear from Figure 2 that there is a strong correlation between the measured temperature data and the satellite data. In order to improve this relationship, the R^2 value was tried to be increased. Accordingly, second degree polynomial equation data was obtained from the satellite data and using the first coefficients of this equation, the ground data was smoothed. As seen in Figure 3, when the saturated temperature and satellite data are regraphed, an improvement in the range of 10% is achieved for air temperature in R^2 value.

For wind speed, equation (4) was developed regardless of weather conditions:

$$W_{smooth} = 0.0087W^2_{satellite} - 1.4167W_{ground} + 186.01 \tag{4}$$

This equation in W_{smooth} smoothed wind speed data, $W_{satellite}$ wind speed in MERRA-2 satellite data, W_{ground} wind speed in ground data.

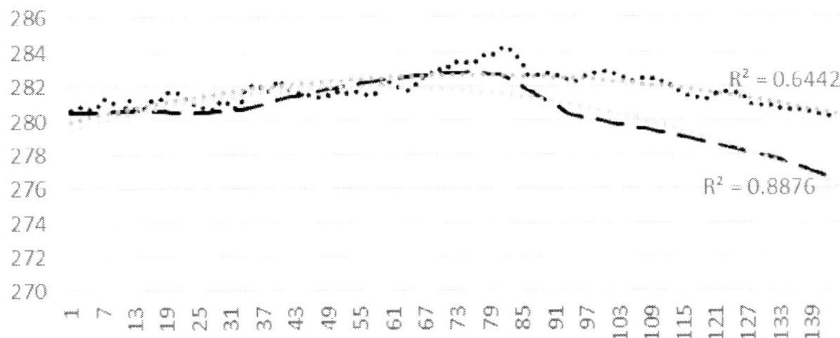

Figure 2. Daily temperature data (dotted-location data, lined-satellite data) average of 10 minutes of weekly data.

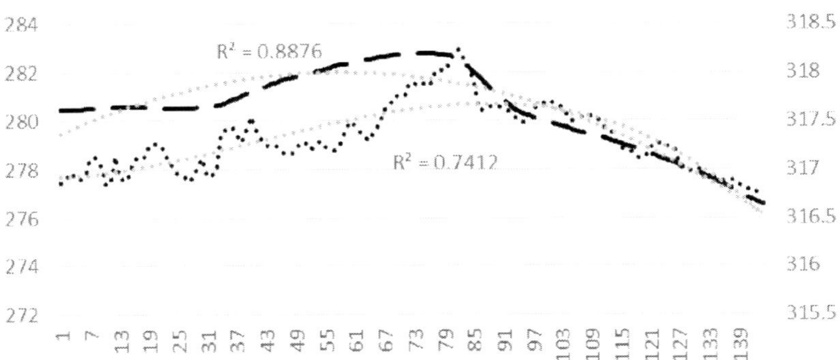

Figure 3. Daily smoothed temperature data analysis. The dotted line is a polynominal fit of the equation and the straight line shows the variation of the difference between the data.

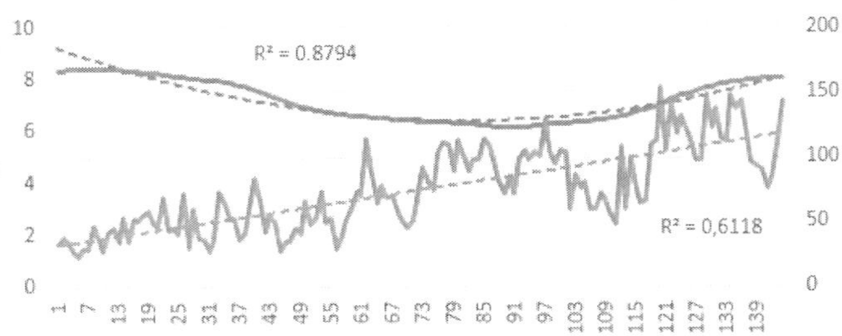

Figure 4. Daily temperature data (dotted-location data, lined-satellite data) average of 10 minutes of weekly data.

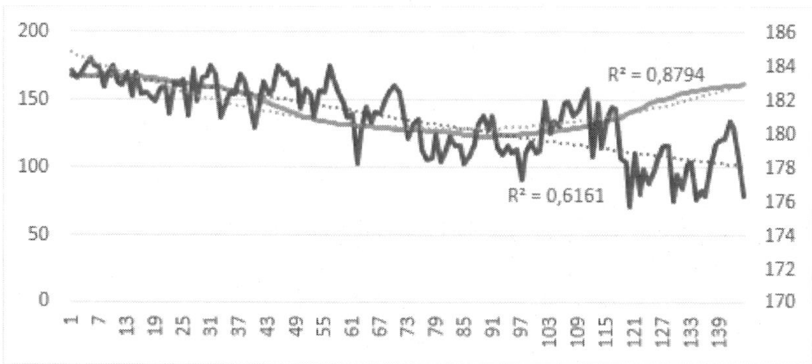

Figure 5. Daily smoothed temperature data analysis. The dotted line is polynominal fit of the equation and the straight line shows the variation of the difference between the data.

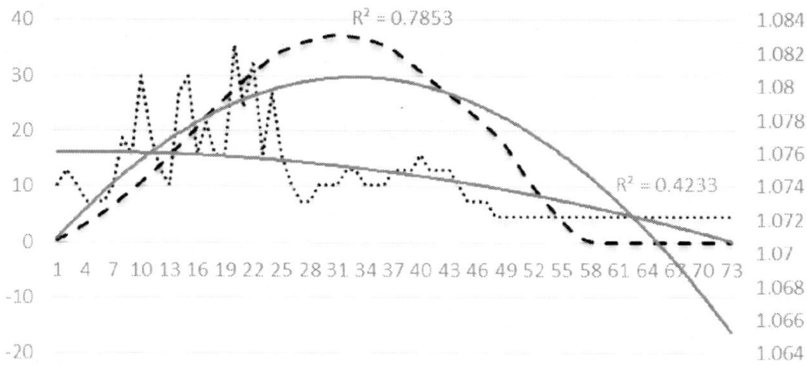

Figure 6. Daily total solar irradiation data (dotted-location data, lined-satellite data) average of 10 minutes weekly data.

As seen in Figure 4, there is a weak correlation between the wind speed data measured from the ground and received from the satellite. One strategy can be to adopt different methods to estimate wind speed. Or to integrate higher sensitivity measurement devices to obtain wind data. As can be seen in Figure 5, there has been an improvement in R^2 value.

For solar irradiance, equation (5) was developed regardless of weather conditions:

$$SI_{smooth} = (-0.028 * SI^2_{satellite}) + (1.8732 * SI_{ground}) + 1.0722 \qquad (5)$$

This equetion in SI_{smooth} smoothed global solar irradiance data, $SI_{satellite}$ global solar irradiance in MERRA-2 satellite data, SI_{ground} global solar irradiance in ground data.

As seen in Figure 6, there is a strong relationship between the measured solar radiation data and satellite data. To further improve the R^2 value in the location data, the second order polynomial equation was used. In this equation, satellite data and ground data are smoothed by using first coefficients. In Figure 7, the compatibility between the two data is shown again. In the smoothed ground data, an improvement in the range of 18% is achieved for solar irradiation in R^2 value. It can be concluded that the prefered approach is for improvement. Therefore, especially when improving the temperature, wind speed and solar irradiance data, the improvement in R^2 values were taken into consideration. If the available data is smoothed, a higher value than R^2 of the previous value was obtained. As an example, 10-minute 1-week data results per day were examined. Because of smoothing the available data, R^2 value increased. That is, location data has become more compatible with satellite data. In order to measure the success of the forecast, the square root of the mean square error (RMSE) value was taken into account and the amount of improvement has been increased as a result of the transaction. R^2 value obtained with improved data has been increased. Therefore, the R^2 value of the data smoothed with the previous R^2 values of the location data has yielded better results.

If the equations resulting from this study can be further developed and the data can be smoothed, instant air measurement results can be obtained with an easy and portable device without the need for location measurement devices. The use of parameters solar irradiation and temperature in wind speed modeling may also increase success. It is considered as an important research topic for future studies. If for wind speed the presented method parameters improvement, it can be used for modeling wind speed data measured from anywhere in the world.

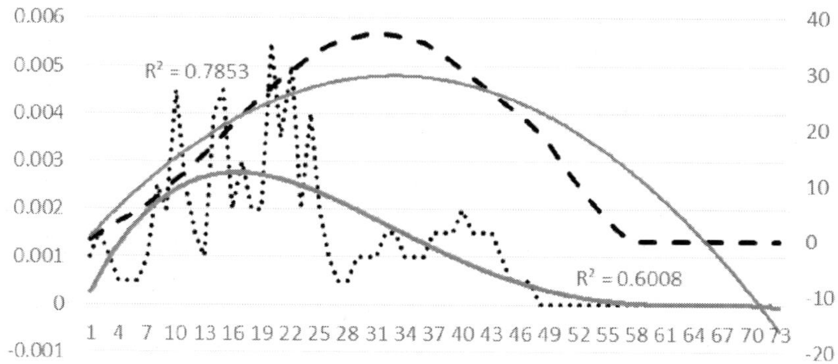

Figure 7. Daily smooted total solar irradiation data analysis. The dotted line is polynomial fit of the equation and the straight line show the variation of the difference between the data.

REFERENCES

[1] Gelaro R. et al., "The Modern-Era Retrospective Analysis for Research and Applications, Version 2 (MERRA-2)," *Journal of Climate*, 2017, https://doi.org/10.1175/JCLI-D-16-0758.1.

[2] Chen X. et al., "Short-Term Wind Speed Forecasting Study and Its Application Using a Hybrid Model Optimized by Cuckoo Search," *Mathematical Problems in Engineering*, 2015, https://doi.org/10.1155/2015/608597.

[3] Menezes E. J. N., Alex Maurício Araújo and Nadège Sophie Bouchonneau da Silva, "A Review on Wind Turbine Control and Its Associated Methods," *Journal of Cleaner Production*, 2018, https://doi.org/10.1016/j.jclepro.2017.10.297.

[4] IRENA, International Renewable Energy Agency. *Renewable Power Generation Costs in 2017*, International Renewable Energy Agency, 2018.

[5] Saleh H., A. Abou El-Azm Aly and S. Abdel-Hady, "Assessment of Different Methods Used to Estimate Weibull Distribution Parameters for Wind Speed in Zafarana Wind Farm, Suez Gulf, Egypt," *Energy*, 2012, https://doi.org/10.1016/j.energy.2012.05.021.

[6] Liang J. et al., "An Open-Source 3D Solar Radiation Model Integrated with a 3D Geographic Information System," *Environmental Modelling and Software*, 2015, https://doi.org/10.1016/j.envsoft.2014.11.019.

[7] Journée M. and Cédric Bertrand, "Improving the Spatio-Temporal Distribution of Surface Solar Radiation Data by Merging Ground and Satellite Measurements," *Remote Sensing of Environment*, 2010, https://doi.org/10.1016/j.rse.2010.06.010.

[8] Whitaker J. C., "Electromagnetic Spectrum," in *The RF Transmission Systems Handbook*, 2017, https://doi.org/10.1201/9781420041132.

[9] Monteith J. L. and M. H. Unsworth, "Principles of Environmental Physics, 2nd Edition.," *Edward Arnold, London*, 1990, https://doi.org/10.1016/B978-0-12-386910-4.00026-3.

[10] Torres J. L. et al., "Forecast of Hourly Average Wind Speed with ARMA Models in Navarre (Spain)," *Solar Energy*, 2005, https://doi.org/10.1016/j.solener.2004.09.013.

[11] Kavasseri R. G. and Krithika Seetharaman, "Day-Ahead Wind Speed Forecasting Using f-ARIMA Models," *Renewable Energy*, 2009, https://doi.org/10.1016/j.renene.2008.09.006.

[12] Erdem E. and Jing Shi, "ARMA Based Approaches for Forecasting the Tuple of Wind Speed and Direction," *Applied Energy*, 2011, https://doi.org/10.1016/j.apenergy.2010.10.031.

[13] Mabel M. C. and E. Fernandez, "Analysis of Wind Power Generation and Prediction Using ANN: A Case Study," *Renewable Energy*, 2008, https://doi.org/10.1016/j.renene.2007.06.013.

[14] Ranganayaki V. and S. N. Deepa, "An Intelligent Ensemble Neural Network Model for Wind Speed Prediction in Renewable Energy Systems," *Scientific World Journal*, 2016, https://doi.org/10.1155/2016/9293529.

[15] Ailliot P. and Valérie Monbet, "Markov-Switching Autoregressive Models for Wind Time Series," *Environmental Modelling and Software*, 2012, https://doi.org/10.1016/j.envsoft.2011.10.011.

[16] Guo Z. H., et al., "A Case Study on a Hybrid Wind Speed Forecasting Method Using BP Neural Network," *Knowledge-Based Systems*, 2011, https://doi.org/10.1016/j.knosys.2011.04.019.

[17] Shi J., Jinmei Guo, and Songtao Zheng, "Evaluation of Hybrid Forecasting Approaches for Wind Speed and Power Generation Time Series," *Renewable and Sustainable Energy Reviews*, 2012, https://doi.org/10.1016/j.rser.2012.02.044.

[18] Zhang W., et al., "Short-Term Wind Speed Forecasting Based on a Hybrid Model," *Applied Soft Computing Journal*, 2013, https://doi.org/10.1016/j.asoc.2013.02.016.

[19] Fazelpour F., Negar Tarashkar and Marc A. Rosen, "Short-Term Wind Speed Forecasting Using Artificial Neural Networks for Tehran, Iran," *International Journal of Energy and Environmental Engineering*, 2016, https://doi.org/10.1007/s40095-016-0220-6.

[20] Guo Z. H., et al., *"A Case Study on a Hybrid Wind Speed Forecasting Method Using BP Neural Network."*

[21] Monteith J. L. and M. H. Unsworth, *"Principles of Environmental Physics, 2nd Edition."*

[22] Angstrom A., "Solar and Terrestrial Radiation. Report to the International Commission for Solar Research on Actinometric Investigations of Solar and Atmospheric Radiation," *Quarterly Journal of the Royal Meteorological Society*, 1924, https://doi.org/10.1002/qj.49705021008.

[23] Campbell G. S. and John M. Norman, *An Introduction to Environmental Biophysics, An Introduction to Environmental Biophysics*, 1998, https://doi.org/10.1007/978-1-4612-1626-1.

[24] Reindl D. T., W. A. Beckman, and J. A. Duffie, "Diffuse Fraction Correlations," *Solar Energy*, 1990, https://doi.org/10.1016/0038-092X(90)90060-P.

[25] Batlles F. J., et al., "Empirical Modeling of Hourly Direct Irradiance by Means of Hourly Global Irradiance," *Energy*, 2000, https://doi.org/10.1016/S0360-5442(00)00007-4.

[26] Zhang W. Y., et al., "Short Term Wind Speed Forecasting for Wind Farms Using an Improved Autoregression Method," in *Proceedings -*

2011 International Conference of Information Technology, Computer Engineering and Management Sciences, ICM 2011, 2011, https://doi.org/10.1109/ICM.2011.269.

[27] Masseran N., "Integrated Approach for the Determination of an Accurate Wind-Speed Distribution Model," *Energy Conversion and Management*, 2018, https://doi.org/10.1016/j.enconman.2018.07.066.

[28] Huxley P. A., "Incoming Solar Radiation and Degree-Hours of Temperature near Kampala, Uganda," *Agricultural Meteorology*, 1973, https://doi.org/10.1016/0002-1571(73)90088-5.

[29] Walter A., "A Relation between Incoming Solar Radiation and Degree-Hours of Temperature," *Agricultural Meteorology*, 1969, https://doi.org/10.1016/0002-1571(69)90011-9.

[30] Liou K. N., An Introduction to Atmospheric Radiation (Google EBook), *International Geophysics*, 2002, https://doi.org/10.1016/S0074-6142(08)60682-8.

[31] Muneer T., C. Gueymard, and H. Kambezidis, "Solar Radiation and Daylight Data," in *Solar Radiation and Daylight Models*, 2004, https://doi.org/10.1016/b978-075065974-1/50016-4.

[32] Field A., Discovering Statistics Using IBM SPSS Statistics, *Statistics*, 2013.

[33] Benesty J., Jingdong Chen, and Yiteng Arden Huang, "Linear Prediction," in *Springer Handbooks*, 2008, https://doi.org/10.1007/978-3-540-49127-9_7.

[34] Wood L. C. and Sven Treitel, "Seismic Signal Processing," *Proceedings of the IEEE*, 1975, https://doi.org/10.1109/PROC.1975.9798.

[35] Hocaoğlu F. O., Ömer Nezih Gerek, and Mehmet Kurban, "2 Boyutlu Bir Güneş Işinim Şiddeti Modeli," [A Novel 2-D Model Approach for the Prediction of Hourly Solar Radiation], in *2008 IEEE 16th Signal Processing, Communication and Applications Conference, SIU*, 2008, https://doi.org/10.1109/SIU.2008.4632560.

[36] Cadenas E. and Wilfrido Rivera, "Wind Speed Forecasting in Three Different Regions of Mexico, Using a Hybrid ARIMA-ANN Model,"

Renewable Energy, 2010, https://doi.org/10.1016/j.renene.2010.04.022.

[37] Wang J., et al., "A Novel Hybrid Approach for Wind Speed Prediction," *Information Sciences*, 2014, https://doi.org/10.1016/j.ins.2014.02.159.

[38] Chang G. W., et al., "An Improved Neural Network-Based Approach for Short-Term Wind Speed and Power Forecast," *Renewable Energy*, 2017, https://doi.org/10.1016/j.renene.2016.12.071.

BIOGRAPHICAL SKETCHES

Nurdan Karapinar

Affiliation: PhD Candidate.

Education: Ankara University of Astronomy and Space Science.

Business Address: Ankara University, Faculty of Science, Dept. of Astronomy and Space Sciences, E Blok, 2. Floor, TR-06100 Tandoğan/Ankara – Turkey.

Research and Professional Experience:
TÜBITAK 3001 Project Scholarship - 2014/2016;
BA, Ege University Astronomy and Space Sciences;
MSc, Akdeniz University, Space Sciences and Technologies;
PhD, Ankara University, Astronomy and Space Science, Astrophysics.

Professional Appointments: Computational astronomy, Data analysis, Solar physics, Astrophysics.

Publications from the Last 3 Years:
Atay T., M. Kaplan, Y. Kilic, N. Karapinar, 7 July 2016, A-Track: A new approach for detection of moving objects in FITS images, *Computer Physics Communications*, 207:524-530.

Numan Sabit Çetin, PhD

Affiliation: Ege University.

Education:

Degree: Associate Professorship.
Department: Electrical and Electronics Engineering.
Date: 20.03.2015.
Authority: ÜAK (Inter-University Committee).

Degree: PhD.
Department: Solar Energy Institute, Energy Technology Department.
Thesis Title: System Optimization of the Autonomous PM Generated Wind Turbines by Artificial Neural Networks.
Graduation Date: 25.09.2006.
Institute: Institute of Science and Technology.
University: Ege University.

Degree: MSc.
Department: Electrical Training Thesis Title: Output Harmonics Suppression and Input Power factor correction of Uninterruptible Power Supply.
Graduation Date: 10.03.1999.
Institute: Institute of Science and Technology.
University: Kocaeli University.

Degree: BSc.
Department: Electrical-Electronic Training.
Department: Electrical Teaching.
Graduation Date: 23.07.1990.
Faculty: Faculty of Technical Education.
University: Marmara University.

Business Address: Ege University Solar Energy Institute EVKA-3, Bornova, 35100, Izmir, Turkey.

Research and Professional Experience:

He worked as a technical teacher in The Ministry of Education for a period of 3 years from 1990-1993. Between the years 1993-2000, he kept working as a Lecturer and head of Electrical and Electronics Engineering Department in Atatürk University. In 1999, he completed his master's thesis "Output Harmonics Suppression and Input Power factor correction of Uninterruptible Power Supply" in Kocaeli University Institute of Science and Technology. Furthermore, between 1997-1998 he served as a visiting scholar Electrical and Electronics Engineering Department at Victoria University (Melbourne/Australia). From 2000-2003 he assumed Celal Bayar University Deputy Director of Kırkağaç Vocational School. In 2006, he completed his PhD education "System Optimization of The Autonomous PM Generated Wind Turbines by Artificial Neural Networks" in Ege University Solar Energy Institute. Between the years of 2003-2007, he worked as a lecturer at Ege University Ege Vocational School. In 2007, he started working as an Assistant Professor at Ege University Solar Energy Institute. At the same time he founded the Wind and Renewable Hybrid Energy Systems R & D Group under the Department of Energy Technology. Since 2010, he is a Board Member of Solar Energy Institute. In 2011, for 3 months, he served as a visiting researcher at The University of Melbourne (Melborne/Australia) Faculty of Science. On 03/20/2015, he received the title of associate professor and authority in the field of Science in Electrical and Electronics Engineering. He is still continuing his academic studies in the Department of Energy Technology, which studies Renewable Energy Basics, Wind Energy Conversion Systems, Wind-PV Hybrid Power Systems, Electrical Machinery, Electrical and Power Electronics. From 2007 to today, it has been graduated 7 (seven) Masters and 8 (eight) PhD students by him. 5 scientists of his doctoral students serve as an assistant professor in various universities. At the same time he carries out his consultancy of 20 master's and doctoral students. In addition, he has served in executive positions for 12 projects, advised on

12 projects and made a researcher for 5 projects within national and international R & D projects (EU, TUBITAK, RDAs, SRP, etc). However, he did refereeing of 25 TUBITAK R & D projects. There are already ongoing and application stages his new projects. In 18 different Scientific Seminar/Conference organizing events, he made a member of the organizing committee, a member of the scientific committee and panelist task. He has published 21 articles in international peer-reviewed journals and 9 articles in national peer-reviewed journals. There are 50 papers presented at national and international meetings.

Honors: 2017- Science-Art Award/Coming from the Future Science and Technology Association.

Publications from the Last 3 Years:

Bawazir R. O., Çetin N. S., "Comprehensive overview of optimizing PV-DG allocation in power system and solar energy resource potential assessments," *Energy Reports,* cilt. 6, ss. 173-208, 2020.

Kandemir E., Çetin N. S., Börekci S., "An Investigation of Intelligent and Conventional Maximum Power Point Tracking Techniques for Uniform Atmospheric Conditions," *European Journal of Engineering and Natural Sciences,* cilt. 3, ss. 93-100, 2019.

Kandemir E., Börekci S., Çetin N. S., "Comparative Analysis of Reduced-Rule Compressed Fuzzy Logic Control and Incremental Conductance MPPT Methods," *Journal of Electronic Materials,* cilt. 47, ss. 4463-4474, 2018.

Çetin N. S., Deniz E., Başaran K., "Maximum Power Point Tracking on the Alternative Current Side forPhotovoltaic Applications," *Afyon Kocatepe University Journal of Science and Engineering,* cilt. 18, ss. 495-503, 2018.

Karabacak K., Çetin N. S., "Design and Application of an ANN Controlled Off Grid Inverter for PV Power Systems," *Modern Environmental Science and Engineering,* cilt. 4, ss. 131-136, 2018.

Kandemir E., Börekci S., Çetin N. S., "Conventional and Soft-Computing Based MPPT Methods Comparisons in Direct and Indirect Modes for

Single Stage PV Systems," *Elektronika Ir Elektrotechnika*, cilt. 24, 2018.

Kandemir E., Çetin N. S., Börekci S., "A comprehensive overview of maximum power extraction methods for PV systems," *Renewable & Sustainable Energy Reviews,* cilt. 78, ss. 93-112, 2017.

Basaran K., Çetin N. S., Börekci S., "Energy management for on-grid and off-grid wind/PV and battery hybrid systems," *IET Renewable Power Generation,* cilt. 11, ss. 642-649, 2017.

Chakchak J., Çetin N. S., "Application of Rural Photovoltaic Water Pumping System Using Immersed Pump and DC Motor," *International Journal of Energy Applications and Technologies,* cilt. 4, ss. 164-173, 2017.

Mwanza M., Kaoma M., Bowa C. K., Çetin N. S., Ülgen K., "The potential of solar energy for sustainable water resource development and averting national social burden in rural areas of Zambia," *Periodicals of Engineering and Natural Sciences,* cilt. 5, ss. 1-7, 2017.

Mwanza M., Chachak J., Çetin N. S., Ülgen K., "Assessment of solar energy source distribution and potential in Zambia," *Periodicals of Engineering and Natural Sciences,* cilt. 5, ss. 103-116, 2017.

Kandemir E., Çetin N. S., Börekci S., "A Comparison of Perturb and Observe and fuzzy-logic based MPPT methods for uniform environment conditions," *Periodicals of Engineering and Natural Sciences,* cilt. 5, ss. 16-23, 2017.

Melih Soner Çeliktaş, PhD

Affiliation: Assoc. Prof. Dr.

Education: PhD, Ege University Solar Energy Institute.

Business Address: Ege University Solar Energy Institute.

Research and Professional Experience:

Melih Soner Celiktas is a mechanical Engineer who graduated from Yildiz Technical University. He obtained his Diploma (MSc) in Mechanical Engineering and his Doctorate (PhD) in Solar Energy Institute, in 2005 and 2009, respectively, both from the Ege University, Turkey. He is working as an associate professor at Ege University Solar Energy Institute. His research interests lies in the field of renewable energy and impacts upon the society and economy. He is currently investigating the biorefinery systems such as Physico-chemical pretreatment, bioconversion of renewable lignocellulosic biomass to biofuel and value added products like biomaterials using the latest conversion techniques and analysis.

Since last decade, He has been working (co-ordinated & collaborated) in a number of EU projects including IRC, FASCAP, EBIC, Posmetrans SPINE and IQVETRES. His research interests include renewable energy, technology foresight, bibliometric analysis, biobased materials, energy economy, renewable energy policy and their implementation in technological and socio-economic fields. His future research plans are to build on the robotics and automation systems engineering on energy crops using biorefinery technics to further develop models and tools in conjunction with other related researchers and industrial actors.

Celiktas has published over 20 research papers in international journals. He has also presented more than 60 papers in national and international meetings and scientific conferences. Celiktas is member of editorial board of scientific Journals and co-author of five books, more than 200 peer-review scientific papers published in international journals and more than 10 book chapters. He has also a national patent on biorefinery technology.

Publications from the Last 3 Years:

Güneş K., Yağlıkci M., Celiktas, M. S., 2020. "Pressurized liquid hot water pretreatment and enzymatic hydrolysis of switchgrass aiming at the enhancement of fermentable sugars" *Biofuels,* doi: 10.1080/ 17597269.2019.1706273, Accepted.

Çeliktaş M. S., Alptekin F. M. 2019. "Conversion of model biomass to carbon-based material with high conductivity by using carbonization." *Energy*, 116089.

Çeliktaş M. S., Yağlıkçı M., Khosravi Maleki F., 2019. "Subcritical water extraction derived lignin for creation of sustainable reinforced composite materials." *Polymer Testing*, 77.

Uyan M., Alptekin, F. M., Bastabak, B., Özgül, S. Erdogan, B., Öğüt, T., Sezer, U. Çeliktaş M. S., 2019. "Combined biofuel production from cotton stalk and seed with a biorefinery approach." *Biomass Conversion and Biorefinery*, doi: https://10.1007/s13399-019-00427-z, Article in Press.

Uyan M., Çeliktaş, M. S., 2019. "Plant Fiber Reinforced Biocomposite: Properties and Applications." *Mugla Journal of Science and Technology*, 5(2), 42-48.

Çebi D., Alptekin, F. M., Uyan, M., Sarptaş, H., Çeliktaş, M. S., 2019. Bioethanol Production from Hazelnut Shell by Liquid Hot Water Pretreatment. *X. International Multidisciplinary Congress of Eurasia.*

Çebi D., Alptekin, F. M., Uyan, M., Sarptaş, H., Çeliktaş, M. S., 2019. "A Review on the Entrepreneurial Structure of Higher Education Institutions Regarding Academic Studies on Energy." *IITEEC 2019 Internat ional Instructional Technologies in Engineer ing Educat ion Conference*, 1(1), 8-8.

Çeliktaş M. S., 2019. "Value added products from bio-based manufacturing and the Future trends of bioconversion and biorefinery Industries." *MAS European International Congress on Mathematics-Engineering-Natural Medical Sciences-III*, 1(1), 255-261.

Çeliktaş M. S., 2019. "Future energy systems based on smart energy infrastructure integratıng renewable energy." *MAS European International Congress on Mathematics-Engineering-NaturalMedical Sciences-III*, 1(1), 262-267.

Gültekin S. Y., Olgun H., Çeliktaş M. S., (2018). "Comparison of solid biofuels produced from olive pomace with two different conversion methods: torrefaction and hydrothermal carbonization." *International*

Journal of Engineering Technology, 7(2.23), 143, doi: 10.14419/ijet. v7i2.23.11903.

Ucar R. C., Sengul, A., Çeliktaş, M. S., 2018. "Sustainable Recovery and Reutilization of Cereal Processing By-Products, Chapter 4. Wheat bran-based biorefinery" (2018), *Elsevier,* Editor: Charis Galanakis, Basım sayısı: 1, Sayfa Sayisi 400, ISBN: 9780081021620.

Alptekin F. M., Çakır M., Çeliktaş, M. S., 2018. "Supercapacitor as an Energy Storage Device: Current And Future Prospect." *SOLARTR, 1,* 45-58.

Uyan M., Çakır M., Çeliktaş, M. S., 2018. "Plant fibre based biocomposites and the future course available." *SOLARTR*, 289-302.

Uyan M., Çeliktaş, M. S., 2018. "Future prospects of biocomposites." *IV. International Ege Composite Materials Symposium, KOMPEGE 2018,* 1(1), 390-400.

Alptekin F. M., Çeliktaş, M. S., 2018. "Nanocomposites and its applications in energyconversion and storage devices." *IV. International Ege Composite Materials Symposium, KOMPEGE 2018,* 1(1), 531-545.

Celiktas M. S., Uyan, M., Alptekin, F. M., 2017. "Biorefinery concept: Current status and future prospects." *International Conference on Engineering Technologies (ICENTE'17),* p. 27-32, Konya, Turkey.

Cerone N., Zimbardi, F., Celiktas, M. S., Vito, V. 2017. "Pilot plant air steam gasification of nut shells for syngas production," presented at *the 25th European Biomass Conference and Exhibition*, pages 843-846.

Deniz E., Celiktas, M. S., 2017. "Overview of future for offshore wind energy in Turkey." *7th International 100 Renewable Energy Conference, IRENEC* 2017, p. 105-110.

In: Wind Speed: An Overview
Editor: Nicolas Koči

ISBN: 978-1-53618-412-9
© 2020 Nova Science Publishers, Inc.

Chapter 4

OVERVIEW OF WIND CONDITIONS AT KOSOVO

Rexhep Selimaj, Bukurije Hoxha and Sabrije Osmanaj*
University of Prishtina, Faculty of Mechanical Engineering,
Sunny Hill, Prishtina, Kosovo

ABSTRACT

Growing problems of global warming, environmental pollution and security of energy generation have increased interest in developing renewable and wind-friendly energy sources for different countries, and Kosovo has started it. This chapter studies wind as a source of energy in Kitka and Koznica. Initially, the study analyses the types of winds in the two potential sites in Kosovo, for which there are real measurements for the purpose of investing in this sector. Aerodynamics is also an important part of the study. In order to harness the wind energy, it is necessary to carry out terrain conditions analyses for the installation of wind turbines, which are supplemented by long-term correlation analyses. From the installed turbines, for their specific capacity and the number of potential places to put them, the annual electricity generation, generated by the wind park is calculated, taking into account the capacity factor of the wind turbines for this terrain. Further the economic analysis for wind park

---

* Corresponding Author's Email: bukurije.hoxha@uni-pr.edu.

installation is given, where we can see what the lifecycle of two possible wind power generation projects can be. Economic analysis shows a life span of approximately 25 years for wind parks, and a return on investment of about 9.2 years, values that vary depending on the trend of selling a kWh, i.e., the price of electricity sold in a country.

Keywords: wind energy, production of electrical energy, efficiency, wake losses, economic analyses

INTRODUCTION

Kosovo has committed itself to recognizing the savings target of 9% of the total consumption of the service, subject to the EU Directive 2006/32/EC. It has also undertaken obligations to implement new EU policies stemming from the European Services Directive 2012/27/EC.

The problem of electricity supply in Kosovo is from the 90th years. Since 1984 there are no new thermal power plants or plants more you can offer smaller capacity like hydro, we as a country started building hydropower plants in a large capacity from 2007. Currently most power plant blocks exist on their tire end.

In order to improve this way of generating energy in Kosovo, it is necessary to cease operating the Kosova A power plant (as possible in the true pursuit of his right) South East Europe (SEE).

In previous years, domestic production and some by imports have overwhelmingly driven electricity demand coverage. In predominantly over-loaded power system situations and/or reasonably priced imports of energy at a reasonable price then there were planned reductions (under the ABC plan). The reductions have been cancelled in recent years due to increased domestic electricity production. Although this plan has not been formally cancelled, it is currently not being implemented (Economic, 2017).

From 2000 to 2015, EUR 538.25 million was spent on electricity imports. In some cases there have been government interventions on import subsidies and since 2012, subsidies on imports have been discontinued. Following the privatization of the company for distribution and supply of electricity, the cost of importing electricity has fallen sharply, inter alia as a result of more efficient procurement procedures.

Renewable Energy Sources (RES) represent an important source of energy available to Kosovo, with a potential that is still underutilized. Utilizing these resources for energy production is a long-term objective for achieving the goals of the country's energy policy such as: supporting overall economic development; increasing the security of energy supply and protecting the environment. In view of these goals, there is a need to apply fiscal and financial incentives for all types of RES, including the implementation of a support scheme based on the certificate of origin system (Naser Sahiti, 2013).

Energy sector laws, in particular the Law on Energy, have consistently addressed Renewable Energy Sources in terms of promoting the optimization of their utilization, including setting annual and long-term targets for energy production from these sources. To support and promote the use of Renewable Energy Sources, the Ministry of Economic Development has also drafted a ten-year RES action plan as a policy document for this important energy sector. In accordance with the legal obligations and those taken under the Energy Community Treaty (ECT), MED has set RES targets for the period 2011-2020, taking into account the potentials of Renewable Energy Sources with which Kosovo has. Completion of these targets is monitored by the Ministry of Economic Development, which is also responsible for reporting progress to the Energy Community Secretariat (ESC) in Vienna (Regulatory, 2018).

The two sites with the highest wind potential in Kosovo are Kitka and Koznica, where Kitka now operates a wind park with a capacity of approximately 32MW (Osmanaj, 2018). But since Koznica is a future wind park that is not built, then the analysis will mainly refer to the future park in Koznica. Capacity of wind power that can be for 10 wind turbines to be considered (Osmanaj, 2018), (Hoxha, 2018).

1. STUDY OF THE TYPE OF WINDS - AERODYNAMICS, AND BASIC EQUIPMENT FOR ITS USE

In Kosovo there is still no data on its entire territory regarding wind potential and wind characteristics, which could be used to dimension potential wind power generation capacities. So Kosovo still has no wind atlas. The data presented below are from the Kosovo Hydro meteorological Institute which conducts continuous measurements at three Kosovo locations in Pristina, Peja and Ferizaj.

Table 1. Wind data for different places, in 2012

Places	Pristina	Peja	Ferizaj
Wind speed (m/s)	1.7	1.3	1.52
Relative humidity (%)	67.4	76.8	73.4

Table 2. Wind data for most of cites in Kosovo

Name of place, based on IPKO measurement	City	Height above sea (m)	Height where anemometers where placed (m)	Annual wind speed (m/s)
BBUT	Lipjan	1055	37	NE[*]
ETEC	Lipjan	733	33	3.6
EBUD	Gjilan	592	33	3.3
BBUD	Therandë	1667	38	7.0
SDUL	Therandë	858	34	4.4
WGJU	Klinë	578	44	3.8
EABR	Abri e Epërme	763	45	4.6
BBZYM	Prizren	658	37	3.4
SSTA	Kaçanik	578	37	4.1
BBZAT	Rahovec	1016	35	NE[*]

In 2010, the NEK Umwelttechnik AG (NEK) made a study on wind resource. The study aimed to provide wind information and identify suitable sites for wind energy development. 10 locations were selected for the installation of metering equipment. The measuring equipment is located in the antenna towers of the IPKO mobile operator. Data on location, altitude, altitude and wind speed are presented in following table. The measurement was made over a period of 13 months (Bañuelos-Ruedas, 2011).

2. TERRAIN CONDITIONS OF WIND PARK SETTING - LONG-TERM CORRELATION ANALYSIS

Ground roughness is determined by the ESA Globcover 2009 database. Using this database and the WAsP Map Editor Software package, a field roughness map was developed that spanned a wider region consisting of (20x20) km of territory of the Koznica region. The terrain on which the Koznica park is planned is relatively simple in terms of roughness and can be described with relatively high roughness wooded valleys where wind turbines are planned.

Figure 1. The roughness map of the wider terrain at the target region of WPP Koznica (which is indicated in red contour line).

The terrain orography (variation in elevation) is crucial for the analysis of local winds and should be modelled as accurately as possible. Accurate modelling of terrain orography is essential for both complex terrain (hilly and mountainous sites) and lowland regions, because any change (even minor) in terrain elevation has a significant impact on airflow. Modelling of terrain orography in the Koznica region is performed using topographic maps created using the SRTM (Shuttle Radar Topography Mission) database. Elevation grids were searched for wells, or other hazardous layers, and elevation contours were constructed using Global Mapper software. Figure 2, shows the topographic map of the Koznica region.

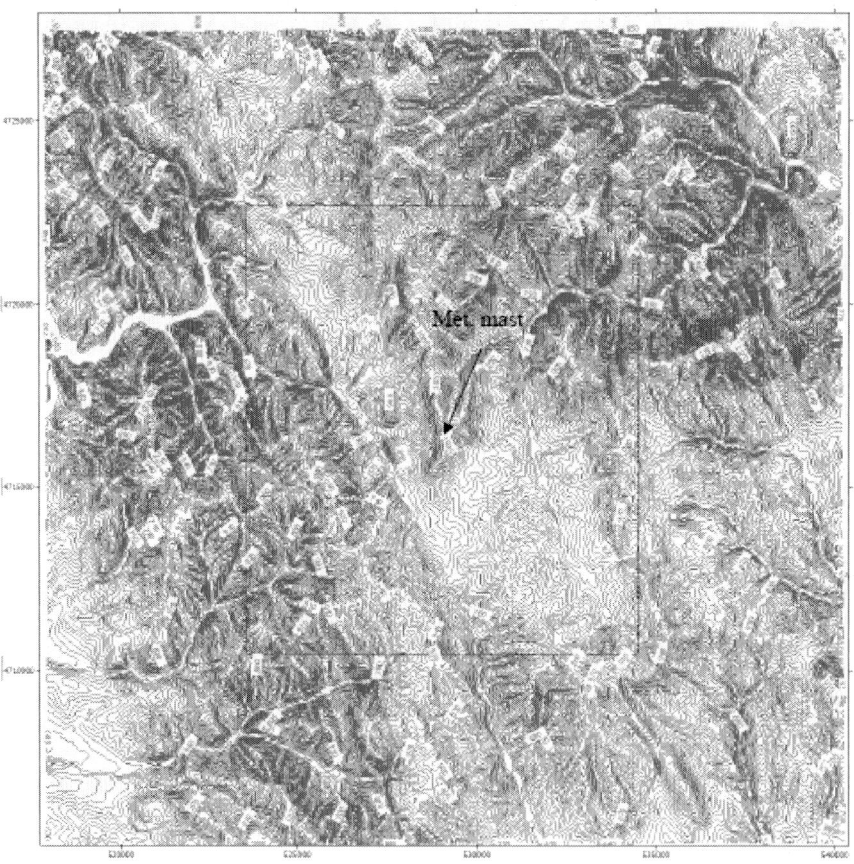

Figure 2. Topographic map of the wither (20x20 km) target region of WPP Koznica.

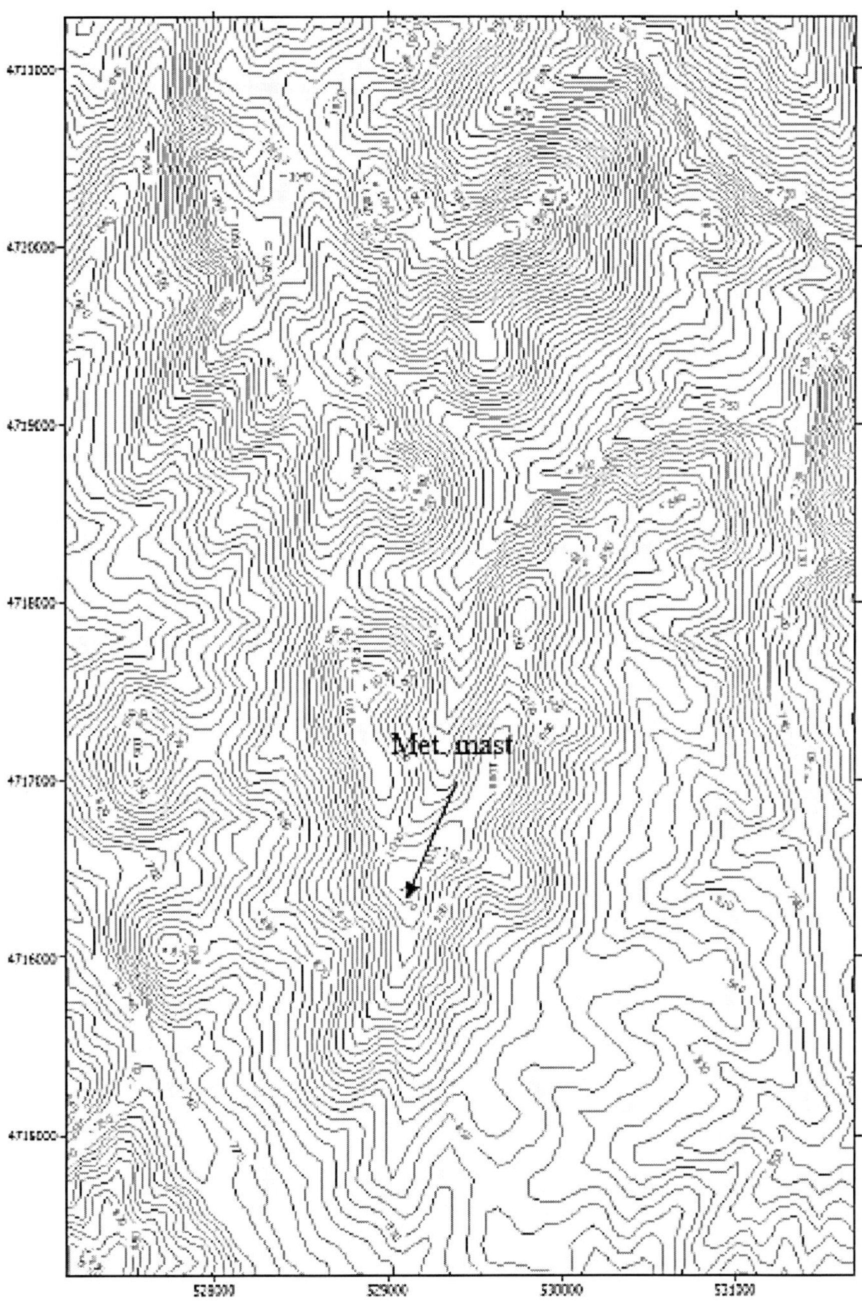

Figure 3. Topographic map of the target region of WPP Koznica.

The kinetic energy of an air current with mass m and moving at a speed w is given by the equation:

$$E_k = \frac{1}{2} \cdot m \cdot w^2 \tag{1}$$

Consider a wind turbine rotor, which has a surface A, exposed to a wind current as shown in Figure 4.

The kinetic energy of the airstream available to the turbine can be expressed as:

ρ_a - air density, in kg/m^3, and

V - volume of air that hits the rotor shafts, in m^3.

The volume of air interacting with the rotor per unit time hits the surface of the A_T turbine rotor, so energy per unit time, that is, power, can be expressed as:

$$P = \frac{E_k}{\tau} = \frac{1}{2} \cdot A_T \cdot \rho_a \cdot w^3 \tag{2}$$

From the equations above, we can see that the factors affecting the available power in the wind flow are air density, wind turbine rotor surface area and wind speed. The effect of wind speed is most noticeable due to its cubic relation to power (Bañuelos-Ruedas 2011).

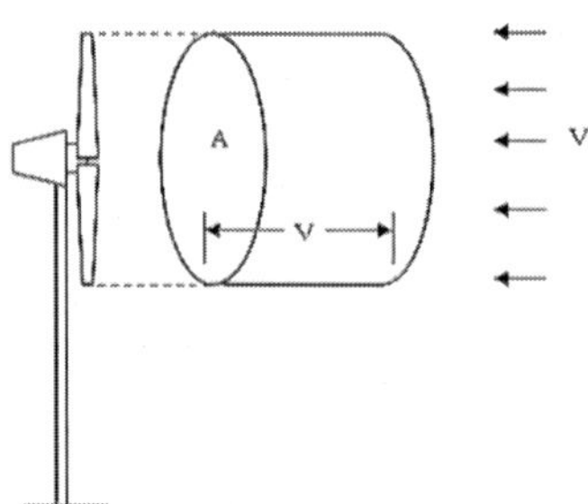

Figure 4. An air mass moving toward a wind turbine.

The main task of a wind park is to obtain as much power as possible from the minimum number of wind turbines and with a minimum space between the turbines due to land economy in land costs and offshore connections. However, minimizing the distance between the turbines within a wind park causes an increase in the so-called oscillation effect, which results from the "shadow" of some wind turbines from other units within the park and consequently causes a turbulent air movement (DNV-GL 2019).

This leads to a reduction in energy efficiency in the shadow units. For offshore wind parks consisting of 100 or more wind turbines this effect is very significant (Uchida, 2017). A typical gear-driven wind turbine consists of multiple components, including blades, main bearing, gearbox, generator, converter, transformer and others (XIn Wu, 2019).

2.1. Long Term Correlation Analyses

Correlations can constrain the line of connections and linearity. At the beginning of a correlation analysis, it is necessary to identify qualitatively whether the correlation is linear or not. Thus, historical wind speeds are treated by dividing wind speeds into several sections, which should ensure that we have a large number of velocity data (Uchida, 2017).

It is a well-known fact that the annual mean wind speed at any given site varies over the years. Therefore, it is important to assess whether the period used for this study is above or below long-term average. A variation of the annual mean wind speed of 5-8% is not unusual, which corresponds to a standard ancertainty of the energy production estimate of about 10-20%. Therefore, in order to assess the long-term wind climate at Koznica site, MERRA data (N42.5; E21.33) covering a period of 37 years (01-01-1979 – 31-12-2015) has been analyzed.

2.2. Long Term Correlation Analyses of Wind Rosse at Target Site Koznica

In order to perform a time series correlation between the measurement data of a reference station and a target station (located at the wind farm Koznica site), the time series of the measured wind data are compared. The relationships of wind speeds and wind directions between them are determined for the common, overlapping measuring period. Afterwards, the correlation parameters obtained by this method will be applied on the long-term (37 years) time-series of the reference series in order to calculate an artificial long-term time-series for the target site. To determine the wind speed relationship a linear regression applied on the wind speed data sets.

Figure 5, shows comparative graphical representation of wind rose on location Koznica obtained from a one year measurement data from the met mast and MERRA data for 37 years.

Following table presents comparative analysis of the frequency of winds for all sectors of the wind rose at Koznica location for one and several year period, corresponiding to the graph in Figure 5.

**Table 3. Comparative analysis of the frequency of occurrence
of wind speed in all sectors of the wind rose
for different sets of measurement data**

	The wind rose sector central angle (deg)											
	0	30	60	90	120	150	180	210	240	270	300	330
	The distribution of wind speed directions (%)											
Koznica wind mast 1 year	3.4	6.1	6.3	3.7	10.6	7.7	3.8	3.3	8.1	11.9	9.3	5.8
MERRA 1 year	6.2	4.1	8.0	5.1	4.6	8.4	7.6	7.7	10.4	7.4	4.3	6.2
MERRA 37 years	5.6	3.1	7.1	4.7	4.6	9.3	8.3	7.5	9.8	8.1	4.8	7.1

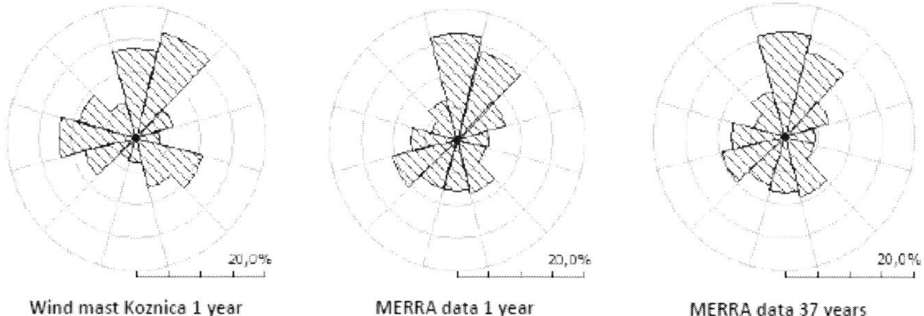

Figure 5. Wind roses at the site Koznica obtained from different sets of measurement data.

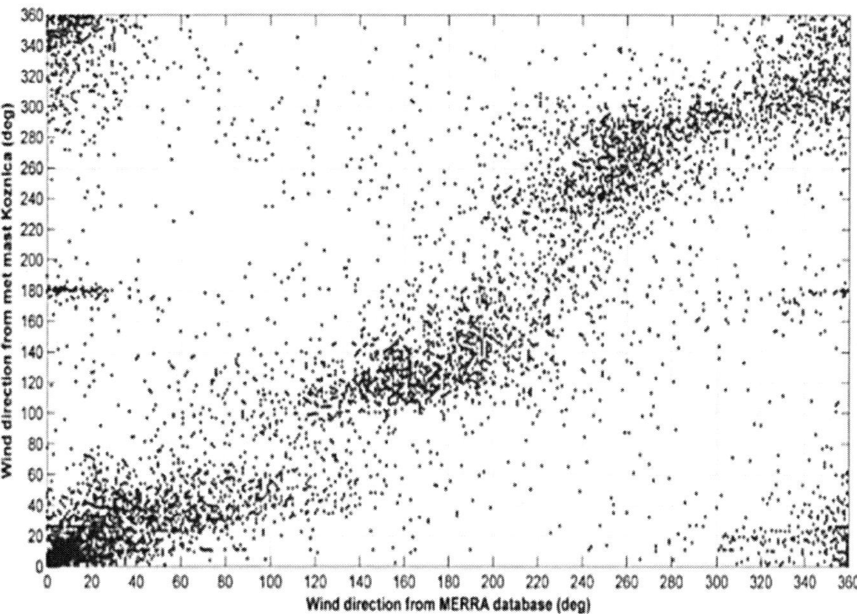

Figure 6. Correlation of measurement data of the wind direction.

From comparative analysis of measurement data for a one-year period and the 37 years period, it can be concluded that the correction of the wind rose is negligible, and that the measured wind rose on the basis of one-year measurement data from the measuring mast may be taken as representative for the design of wind farm layout. Figure 6, shows a cross display of measurement data of wind direction based on one-year measurement data from the met mast at the location Koznica and a time corresponding MERRA data. It can be concluded that dominant winds are time correlated and that belong to the same wind climatology.

Graphical representation of the change of wind direction along 360°, in measurements made in Koznica during one year and those obtained from the MERRA database.

3. ANNUAL ELECTRICITY GENERATION CALCULATIONS

When installing a wind park, it is known that the main purpose is to generate electricity from that park. Even more important than that turbine, specifically the turbine park will be able to operate throughout the year. To calculate this, it is necessary to know the capacity factor of the particular type of turbine depending on the type of turbine that is intended to be installed.

Based on the various manufacturers currently on the market for wind turbines, and in their specifics showing the capacity factor, we can conclude that it ranges from 35-40% as a maximum value, therefore an average value is taken into account when calculating the capacity factor in the annual energy production, as follows:

$$E_{gross} = n_{turbines} \cdot P_{spec} \cdot CF \cdot 8760h / year \tag{3}$$

Based on studies, here we can see that wind potential for a specific capacity is between 3-3.5MW for a wind turbine, an as we assume in beginning if there are 10 wind turbines, and with a specific power capacity 3.3MW, we can calculate annual energy production as follow:

$$E_{gross} = n_{turbines} \cdot P_{spec} \cdot CF \cdot 8760h / year = 10 \cdot 3.3MW \cdot 0.36 \cdot 8760h / year = 104068.8MWh \tag{4}$$

There are acceptable some challenges, for example capacity factor during day, during night, and we can construct a diagram where are shown relationships between gross energy produced and capacity factor.

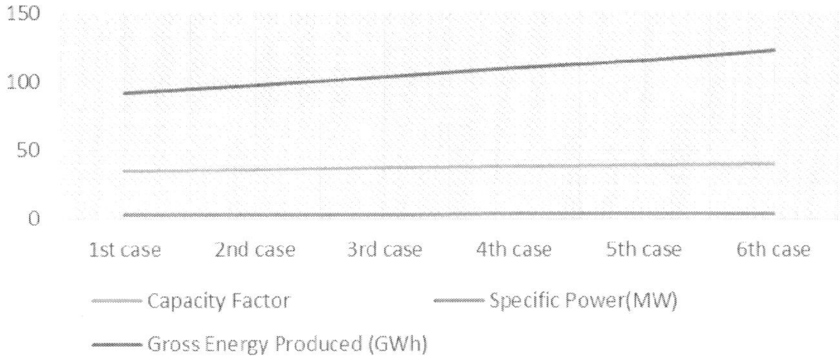

Figure 7. Relationships between capacity factor, specific power and gross energy produced during year.

Wind turbines are mainly clustered as wind farms. Due to the limited space, the turbines are densely placed to obtain the full potential of the available space and to avoid unnecessary cabling costs especially at offshore sites (Martin Cardun, 2019).

4. ECONOMIC ANALYSES FOR KOZNICA WIND FARM

A modern wind turbine will be designed to operate for 120,000 hours over its lifespan of approximately 20 years. The turbine would have to operate at around 66% for two decades. This is, therefore, more than a normal engine which is supposed to last for 4000 to 6000 hours in use. From the experience in Denmark, the cost of maintaining a new wind turbine is much less than that of an old wind turbine. The offshore wind turbine can stay longer than on land, for the simple reason that it does not face offshore wind barriers and the turbulence is much lower. In fact, this would result in a low maintenance cost, but this would be offset by costs in order to reach the inshore turbines and do any maintenance activity.

In terms of facilities to be secured and maintained, it is imperative that these equipment's have maintenance workers, cleaners, security guards and other technicians available under all circumstances. Maintenance and caregivers are responsible for cleaning but also ensuring people in the vicinity, in other words they have to make sure that there are no outsiders on site.

Maintenance workers are responsible for the machinery part, they are tasked with keeping a turbine operating in safe conditions and any damage can be repaired. Wind turbine technicians are responsible for keeping the turbine running in a proper manner, those technicians climbing up and down the tower to make sure the shovels are upright and windward. When there is a problem, those technicians should be available in order to fix it as soon as possible. To build and maintain a turbine requires a large number of technicians, technicians, working in wind parks on a daily basis. Each of these workers along the supply chain contributes to the creation of a viable energy source.

Otherwise this rate of return indicates that Capital Investments in this Project will generate an accumulation with an average annual rate of 13.80%, even this is the limit of the rate allowed for any eventual loan that this project can afford, in the event that all investments would be financed by credit.

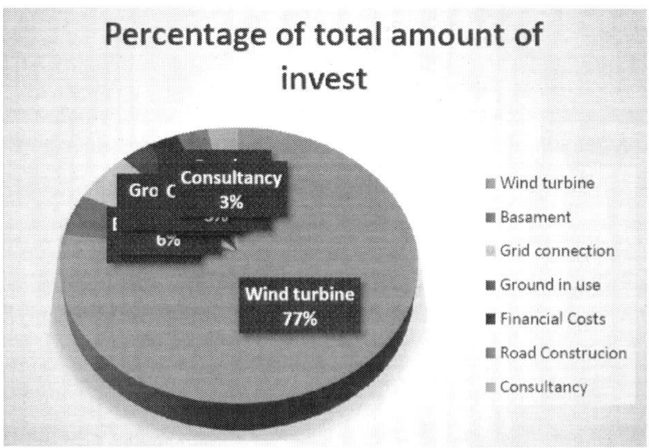

Figure 8. Percentage of components of the wind turbine in total cost graphically.

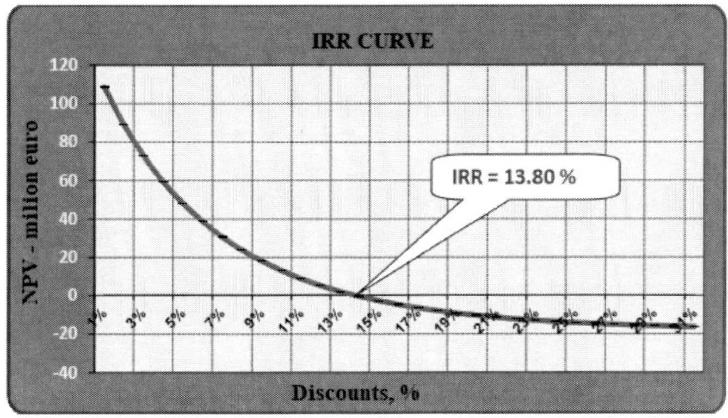

Figure 9. Interior return diagram IRR.

The use of wind power has been developing rapidly all over the world during the past few decades (Xiaodong Wang, 2019). The sensitivity analysis that considers the risks described that the capital investment cost and O & M cost is sensiti ve to changes. Therefore, these conditions affect the change. Meanwhile equity portion, debt portion and loan rate are not sensitive so that they do not affect the change (Kamal, 2015).

CONCLUSION

From the above analysis, it follows that Kosovo, namely the two countries that have been identified as having potential of about 30MW and beyond. In addition to the physical and technical potential of wind in a country, economic analyses of wind parks are also needed. An IRR value of about 13.8 indicates the economic viability of the wind park, which is further complemented by the analysis of the selling price of electricity produced. Kosovo, a small country but with large wind energy capacity. In addition to the two sites that were considered high capacity sites, also the measurements made by a mobile operator, not very high in relation to the land. Beside the losses, it is necessary to examine the representativeness of the input data of wind speed for the evaluation of multi-year production of wind turbines Koznica. Therefore, two corrections connected to production are respected, including the correction due to systematic errors that can be made by measurement and correction due to the long term variations of wind potential.

REFERENCES

Bañuelos-Ruedas, Francisco, César Ángeles Camacho, and Sebastián Rios-Marcuello. 2011. "Methodologies used in the extrapolation of wind speed data at different heights and its impact in the wind energy resource assessment in a region." In *Wind Farm - Technical Regulations, Potential Estimation and Siting Assessment.*

Cardaun, Martin, Roscher, Bjorn, Schelenz, Ralf, Jacobs, Georg. 2019. "Analysis of Wind-Turbine Main Bearing Loads Due to Constant Yaw Misalignments over a 20 Years Timespan." *Energies.*

DNV-GL. 2019. *"Power and renewables."*

Hoxha, Bukurije, Rexhep Selimaj, and Sabrije Osmanaj. 2018. "An Experimental Study of Weibull and Rayleigh Distribution Functions of Wind Speeds in Kosovo." *Telkomnika.*

Kamal, Samsul, and Budi Hartono. 2015. "Economic feasibility of wind farm: a case study for coastal area in South purworejo." *Energy Procedia* 146-154.

Ministry of Economic 2017. *Energy Strategy of Kosovo 2017-2026.* Prishtina: Ministry of Economic Development.

Osmanaj, Sabrije, Bukurie Hoxha, and Rexhep Selimaj. 2018. "An experimental study of Wind Data of a Wind Farm in Kosovo." *Przegląd Elektrotechniczny* (Przegląd Elektrotechniczny).

Regulatory Energy 2018. *"Annual balance of Electrical and Thermal Energy for."* Prishtina.

Sahiti, Naser, Pireci, Maliq, Veselaj, Besim. 2013. *"Manuali për Burimet e Ripërtëritshme të Energjisë."* Prishtine.

Uchida, Takanori. 2017. "High-Resolution LES of Terrain-Induced Turbulence around Wind Turbine Generators by Using Turbulent Inflow Boundary Conditions." *Open Journal of Fluid Dynamics* 511-524.

Xiaodong Wang, Yunong Liu, Luyao Wang, Lin Ding, Hui Hu. 2019. "Numerical Study of Nacelle Wind Speed Characteristics of a Horizontal Axis Wind Turbine under Time-Varying Flow." *Energies.*

XIn Wu, Hong Wang, Guoqian Jiang, Ping XIe, Xiaoli Li. 2019. "Monitoring Wind Turbine Gearbox with Echo State Network Modeling and Dynamic Threshold Using SCADA Vibration Data." *Energies.*

BIOGRAPHICAL SKETCHES

Rexhep Selimaj, PhD

Affiliation: University of Prishtina.

Education: PhD of Mechanical Engineering.

Business Address: University of Prishtina, Kosovo.

Research and Professional Experience: Renewable Energy, Thermal Energy.

Publications on Three Last Years:
1. "An Experimental Study of Wind Data of a Wind Farm in Kosovo," Sabrije Osmanaj, Bukurije Hoxha, Rexhep Selimaj, Link: http://pe.org.pl/articles/2018/7/5.pdf. *Przegląd Elektrotechniczny,* 2018, (Indexed by Web of Science and Scopus).
2. "An Experimental Study of Weibull and Rayleigh Distribution Functions of Wind Speeds in Kosovo," Bukurije Hoxha, Rexhep Selimaj, Sabrije Osmanaj, Link: http://journal.uad.ac.id/index.php/TELKOMNIKA/article/view/10260/pdf_748, *Telkomnika,* 2018, (Indexed by Scopus).
3. "Cogeneration Of Energy In Solar Systems - A Study Case, Kosovo," Bukurije Hoxha, Rexhep Selimaj, Drenusha Krasniqi, Sabrije

Osmanaj[*], Link: http://ijpeds.iaescore.com/index.php/IJPEDS/article/view/18394/12895. *International Journal of Power Electronics and Drive Systems,* 2019, (Indexed by Scopus).

4. Thermal dynamic analysis of parallel and counter flow heat exchangers, Drenusha Krasniqi, Rexhep Selimaj, Marigona Krasniqi, Risto Filkoski, 2018, *International Journal of Mechanical Engineering and Technology.* https://www.scopus.com/authid/detail.uri?authorId=36619946100.

5. An experimental study of passive methods for islanding detection and protection system of the inverter operation [Eksperymentalna analiza pasywnej metody wykrywania pracy wyspowej i zabezpieczenia falownika]; Osmanaj, S., Sylejmani, B., Limani, M., Selimaj, R. 2018 *Przeglad Elektrotechniczny* https://www.scopus.com/authid/detail.uri?authorId=36619946100.

6. Modeling, regulation, and optimization of thermal flux and temperature to a radiation shield between the two planar surfaces; Berisha, X., Selimaj, R.; 2017, *International Journal of Mechanical Engineering and Technology,* https://www.scopus.com/authid/detail.uri?authorId=36619946100.

7. Modelling, regulation, and optimization of mixture air temperature, indoor air and of the rate of heat flow; Selimaj, R., Berisha, X.; 2017, *International Journal of Mechanical Engineering and Technology,* https://www.scopus.com/authid/detail.uri?authorId=36619946100.

8. Modeling of indoor air temperature depending of heat convection and accumulation in the flat wall; Berisha, X., Selimaj, R.; 2017, *International Journal of Mechanical Engineering and Technology,* https://www.scopus.com/authid/detail.uri?authorId=36619946100

9. Dynamic analysis of plane wall heat storage; Selimaj, R., Berisha, X.; 2017, *International Journal of Mechanical Engineering and Technology,* https://www.scopus.com/authid/detail.uri?authorId=36619946100.

Bukurije Hoxha

Affiliation: University of Prishtina.

Education: PhD Candidate; Master of Mechanical Engineering.

Business Address: University of Prishtina, Kosovo.

Research and Professional Experience: Renewable Energy.

Publications on Three Last Years:

1. "Efficiency Analyses for Small Hydro Power Plant with Francis Turbine," Xhevat Berisha, Bukurije Hoxha, Drilon Meha, Link: https://www.ijmter.com/published-papers/volume-4/issue-10/ efficiency-analyses-for-small-hydro-power-plant-with-francis-turbine/. *International Journal of Modern Trends in Engineering and Research* (IJMTER), 2017 (Indexed by Google Scholar).

2. "An Experimental Study of Wind Data of a Wind Farm in Kosovo," Sabrije Osmanaj, Bukurije Hoxha, Rexhep Selimaj, Link: http:// pe.org.pl/articles/2018/7/5.pdf. *Przegląd Lektrotechniczny*, 2018, (Indexed by Web of Science and Scopus).

3. "An Experimental Study of Weibull and Rayleigh Distribution Functions of Wind Speeds in Kosovo," Bukurije Hoxha, Rexhep Selimaj, Sabrije Osmanaj, Link: http://journal.uad.ac.id/index.php/ TELKOMNIKA/article/view/10260/pdf_748. *Telkomnika*, 2018, (Indexed by Scopus).

4. "Cogeneration Of Energy In Solar Systems - A Study Case, Kosovo," Bukurije Hoxha, Rexhep Selimaj, Drenusha Krasniqi, Sabrije Osmanaj[*], Link: http://ijpeds.iaescore.com/index.php/IJPEDS/article/ view/18394/12895. *International Journal of Power Electronics and Drive Systems*, 2019, (Indexed by Scopus).

5. *"Integration of Wind Energy in the Electric System in Kosovo: Current Situation and Challenges"* Sabrije Osmanaj 1, Bukurije

Hoxha 2[*], Link: Accepted to be published on February 2020. (Indexed by Scopus).

Sabrije Osmanaj, PhD

Affiliation: University of Prishtina, Kosovo.

Education: PhD of Electronic Engineering.

Business Address: University of Prishtina, Kosovo.

Research and Professional Experience: Renewable Energy, Electronics.

Publications on Three Last Years:

1. "An Experimental Study of Wind Data of a Wind Farm in Kosovo," Sabrije Osmanaj, Bukurije Hoxha, Rexhep Selimaj, Link: http://pe.org.pl/articles/2018/7/5.pdf. *Przegląd Elektrotechniczny*, 2018, (Indexed by Web of Science and Scopus.

2. "An Experimental Study of Weibull and Rayleigh Distribution Functions of Wind Speeds in Kosovo," Bukurije Hoxha, Rexhep Selimaj, Sabrije Osmanaj, Link: http://journal.uad.ac.id/index.php/TELKOMNIKA/article/view/10260/pdf_748. *Telkomnika*, 2018, (Indexed by Scopus).

3. "Cogeneration of Energy in Solar Systems - A Study Case, Kosovo," Bukurije Hoxha, Rexhep Selimaj, Drenusha Krasniqi, Sabrije Osmanaj[*] Link: http://ijpeds.iaescore.com/index.php/IJPEDS/article/view/18394/12895. *International Journal of Power Electronics and Drive Systems,* 2019, (Indexed by Scopus).

4. *"Integration of Wind Energy in the Electric System in Kosovo: Current Situation and Challenges"* Sabrije Osmanaj 1, Bukurije Hoxha 2[*], Link: Accepted to be published on February 2020. (Indexed by Scopus).

5. "An experimental study for the islanding detection by the harmonic distortion method and protection system of the inverter," Osmanaj, S., Limani, M., Sylejmani, B. *Przeglad Elektrotechniczny*, ISSN 0033-2097, R. 95 NR 4/2019. doi:10.15199/48.2019.04.44 Link: http://pe.org.pl/articles /2019/4/44.pdf.

6. An experimental study of passive methods for islanding detection and protection system of the inverter operation [Eksperymentalna analiza pasywnej metody wykrywania pracy wyspowej i zabezpieczenia falownika]; Osmanaj, S., Sylejmani, B., Limani, M., Selimaj, R. 2018 *Przeglad Elektrotechniczny,* https:// www.scopus.com/authid/detail.uri?authorId=36619946100.

7. The sensitivity of the hoist system in crane applications from speed control methods at induction motor [Die Empfindlichkeit des Hebesystems von Krananwendungen bei Geschwindigkeitsregelverfahren am Induktionsmotor]; Osmanaj, S., Simnica Aliu, K., Limani, M., Kabashi, Q.; 2018, *Elektrotechnik und Informationstechnik,* https://www.scopus.com/authid/detail.uri? authorId=56069810300.

8. The effect of bandwidth on speech intelligibility in albanian language by using multimedia applications like Skype and Viber, Osmanaj, S., Shala, A., Prevalla, B.; 2017, *International Journal of Electrical and Computer Engineering,* https://www.scopus.com/authid/detail.uri? authorId=56069810300.

INDEX